BOTANY
MADE SIMPLE

BOTANY
MADE SIMPLE

VICTOR A. GREULACH, Ph.D.

Professor of Botany, The University of North Carolina

DRAWINGS BY CHARLES C. CLARE, Jr.

The New York Botanical Garden

MADE SIMPLE BOOKS
DOUBLEDAY & COMPANY, INC.
GARDEN CITY, NEW YORK

ABOUT THIS BOOK

This book represents an effort to survey the science of botany in a simple and non-technical way so that it will be understandable by the reader who has had a minimum of formal training in the subject, and yet in great enough detail so that it is more than just a skeleton outline. The aim has been to include botanical facts and principles that could reasonably be expected to constitute part of the general store of knowledge of any educated citizen, and to omit technical details that are of interest and importance only to professional botanists. Where detailed facts are included they are designed to provide the foundation for important general principles and concepts. Every effort has been made to make the book as up-to-date as possible, and many of the more important advances of recent research are included.

If your only knowledge of botany has been the relatively small amount included in high-school biology courses or introductory college biology courses (as is true for perhaps most people), the reading of *Botany Made Simple* may surprise you pleasantly with respect to the nature and scope of modern botany. Indeed, you may find botany much more interesting than you had supposed it would be. You may be surprised, too, to learn that botany is not just a matter of classifying and identifying plants and that horticulture and agriculture, though they are applied botanical sciences, are not really the core of the science of botany. In any event, if this book brings to you a better understanding and appreciation of plants as a most important part of the world in which we live it will have achieved its goal.

It is our hope that your reading of this book will be only the beginning of your interest in botany and that you will want to go on and read some of the references listed at the end of the book, and perhaps also to engage in one of the botanical hobbies that have proved to be fascinating to many people.

Although this book has been written so that it should be comprehensible without any previous knowledge of botany or other biological sciences, it has been impossible to write Chapter 7 and to a lesser extent Chapters 6 and 13 without assuming at least an elementary knowledge of chemistry on the part of the reader. Those who lack such knowledge will profit by the reading of *Chemistry Made Simple* or a comparable introduction to chemistry before undertaking these chapters. However, even with no knowledge of chemistry, the general principles and concepts in these chapters should be understandable although the chemical aspects may remain rather vague.

—Victor A. Greulach

CONTENTS

8   *Contents*

Chapter 1

FIRST FAMILIES OF THE PLANT KINGDOM

Like other members of the animal kingdom, we are utterly dependent on plants for two major biological necessities—the food we eat and the oxygen in the air we breathe. In addition, we depend on plants for hundreds of our most important and essential items of commerce: clothing, shelter, fuel, and paper. The 350,000 or so species of plants inhabiting the earth belong to almost a thousand different families, but only a few hundred species belonging to a few dozen families have been of any substantial economic importance to man. Indeed, there are ten families of plants whose value to our major civilizations exceeds that of all other plant families combined.

Each of these ten families provides products worth at least a billion dollars a year, and their combined annual value in the United States alone is well over thirty-five billion dollars. From an economic standpoint these are the first families of the plant kingdom. The total value of products from the second ten families in the United States would probably not exceed four billion dollars, despite the fact that they include some of our important and widely used economic plants.

The first ten families are characterized not only by their immense value, but also by the number and diversity of the economic plants in these families and the essential character of many of the products. Selection of the ten families has been limited to present-day plants that are of direct economic importance, thus eliminating from consideration those very ancient plants from which our fossil fuels, such as coal, petroleum, and gas, were formed and the marine algae which provide the basic food supply of ocean animals. For tropical civilizations a list of the first ten families would be quite different from our list, and would probably be headed by the palm family, which just fails to make our list but has over eight hundred uses in the tropics.

We are using the term *family* in its biological, rather than its popular, sense and the difference between the two uses is worthy of note. The popular use of the term is a much more limited one, generally corresponding to the *genus* of biologists. When a layman speaks of the lily family he probably means just the different species and varieties of lilies, while botanically the lily family includes also such plants as tulips, hyacinths, trilliums, yucca, asparagus, onions, and garlic. It does not include the spider lily (*Hymenocallis*), which is not really a lily at all but a member of the amaryllis family. Common names can be misleading.

Now for our ten first families, listed more or less in the order of their importance:

Grass Family (*Gramineae*). In the United States alone products of the grass family have an annual farm value of over fifteen billion dollars, a figure equal to that of the next five families combined. Since grasses are mostly food plants their basic biological importance to man cannot be reckoned in mere dollars and cents. Some grass, such as wheat, rice, or corn, has always been basic in the food economy of each civilization. In America we have a double dependence on the wheat of Europe and the corn (or more properly maize) of the American Indian, though the five-billion-dollar annual farm value of corn is about double that of wheat. Corn is a particularly versatile plant. Both the grains and stalks are important feeds for domestic animals, and the grains provide a wide variety of human foods: cereals, corn meal, hominy, grits, cornstarch, corn oil, corn syrup and sugars, popcorn, and roasting ears. Corn provides us with seventy-five times as much starch as all other plants combined. In addition, corn is the source of an ever-increasing number of industrial chemicals, even the cobs serving as sources.

Among the other important cereal grasses are rice, barley, oats, rye, and grain sorghum. One extensive use of the various grains is in the manufacture of alcohol and alcoholic beverages. However, next to corn and wheat, the most important grasses in our agricultural economy are the various hay and pasture grasses such as timothy, fescue, and Sudan grass. Of course, we must not forget our lawn grasses even though they rank relatively low in total economic value. There are several major grains, such as the delicious wild rice and Job's tears, the latter being raised only as an ornamental in this country although it is used as food in the Far East. Several grasses have minor, but unusual, uses. For example, *Stipa tenacissum* of Africa is the source of esparto wax and *Cymbopogon* of Central America is the source of lemon grass oil, which is used both as a base for synthetic violet perfume and as a source of vitamin A.

Three important grasses deserve special mention: sugar cane, sorghum, and bamboo. Sugar cane is, of course, the world's most important source of sugar, not to mention molasses and rum. Sorghum, like corn, is a most versatile plant, the source of grain, fodder, syrup, broomstraw, and a variety of chemicals. Bamboo is relatively unimportant in our economy, but has ranked next to rice in its importance in Asia. Jeanette Speider, the writer, has pointed out in an article in *Nature Magazine* that in China "One may eat with bamboo chopsticks sitting in a bamboo chair before a bamboo table in a bamboo house. One may travel in comfort lying on a bamboo mat under a screen of woven bamboo while a boatman pushes his craft along with a bamboo pole, shouting and whistling now and then for a wind to come and fill the great bamboo sail. In the streets coolies stride by with bamboo carrying poles filled with water, swerving to avoid the bamboo sedan chairs in which the wealthy ride." She might have added that the menu probably included bamboo sprouts cut with a bamboo knife, and that paper, hairpins, fans, and phonograph needles have also been made from bamboo. It is likely, however, that as time goes on the Chinese will become less and less dependent on this large grass.

Pine Family (*Pinaceae*). The number two family, with an annual American value of over five billion dollars, provides us with products quite unlike those of the grass family. The pine family is insignificant as a source of food, even though the edible seeds such as the piñon nuts of our West and the pignola nuts of Europe are most delicious. The family is important as the source of most of our lumber and

paper, as well as resin and turpentine, Christmas trees, and ornamentals. The cellulose and lignin of the wood are the raw materials used in the manufacture of many important materials such as rayon, celluloid, furfural, alcohols, acetone, and lactic acid. The pine family is the source of 75 per cent of all our lumber and 93 per cent of our softwood or coniferous lumber. While the various pines and the Douglas firs are the most important sources of lumber, the other genera of the family—white fir, spruce, and hemlock—are also important lumber trees.

While we are quite aware of the importance of lumber, we sometimes fail to appreciate fully the complete dependence of our civilization on paper. Without it modern schools, business, government, publications, communications, and libraries would not exist, and even our homes would be greatly handicapped. Without the pine family our supply of paper would be both very limited and very expensive.

Mallow Family (*Malvaceae*). The third family in economic value—unlike others which include numerous important plants—owes its rank to one member, the cotton plant (Fig. 1). Though there are

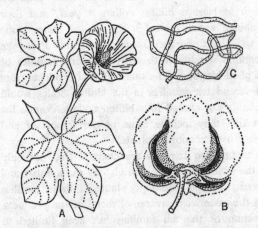

Fig. 1. Cotton. **A**–Flower and leaves. **B**–Cotton boll. **C**–A single cotton fiber as seen under the microscope.

other economic plants in this family they are of relatively minor importance. Several ornamentals—hollyhocks, mallows, hibiscus, and rose of Sharon—belong to the family, as does the vegetable okra (*Hibiscus esculentus*). Amberette, an exalting agent in fine perfumes, is secured from *Hibiscus abelmoschus*. The family is the source of several other fibers besides cotton, such as urena jute, Cuba jute, and China jute. Cotton is not only our leading textile fiber, but the cotton plant is also important as the source of cottonseed oil and cottonseed meal. In addition, cotton linters are used in the manufacture of rayon.

Pea Family (*Leguminosae*). Next to the grasses, the legumes provide the most important source of forage for our domesticated animals and they are also one of the most important sources of human food. The forage plants include such valuable species as the clovers, alfalfa, lespedeza, vetch, lupine, soybeans, and cow peas. For human food there are the various kinds of beans, peas, and peanuts, all of which have high protein contents. Primarily a family of plants useful for food, various legumes have other uses. Indigo dye, once so important but now largely replaced by coal tar dyes, comes from a legume, as does the dark purple logwood dye or haematoxylon. Several useful gums, such as arabic (or acacia), tragacanth, carob, and guaiac, are extracted from legumes, as is the aromatic resin, balsam of Peru. This substance is used in medicines, as a perfume fixative, as a vanilla substitute, and in soaps and consecrated oils. Perfume oils from legumes include acacia, genet, cassie, and mimosa. Coumarin, from the Tonka bean, is used as a tobacco flavoring and a vanilla substitute, while fenugreek is used in making artificial maple flavoring. Licorice is made from the dried rhizomes and roots of *Glycyrrhiza glabra* of the Middle East. Senna is used in cathartics, while the derris plant is the source of an important and effective insecticide, rotenone. Returning to foods, we may mention the tubers of the ground nut, the commercially important peanut and soybean oils, and tamarind. The latter is an Indian plant and one of the few legumes in which the fruit, in the botanical sense, is eaten as a fruit, in the popular sense. It contains more sugar and acid than any other fruit and is an important market item in Arabic countries and the East Indies.

Many legumes are important as ornamental plants. Among these are sweet peas, scarlet runner beans, bluebonnets, mimosa, acacia, redbud, locust, and wisteria. Legumes are of incalculable value in increasing soil fertility, and it is for this reason that farmers always include a legume in their crop rotation plans.

The legume of the greatest economic importance is the soybean. The annual farm value of this plant in the United States alone approaches a billion dollars. Though an important cultivated plant in the Orient for over five thousand years it was not brought to this country until 1804 and has become an important crop here only since 1900. Soybean meal is an important animal feed. Since soybeans contain little sugar and practically no starch they are an ideal source of foods for diabetics. They have three times the protein content of wheat or eggs and twice as much as lean meat, and can be made into cheese, soups, butter, macaroni, breakfast food, and even imitation milk. They are important sources of vitamins, as well as a vast array of non-edible products such as glues, paints, plastics, waxes, axle grease, paper sizing, explosives, linoleum, printing ink, rubber substitutes, soaps, and glycerine. It is said that the Japanese used to refer to the soybean as the "little honorable plant" and would tip their hats when passing it. Today, we might well treat this versatile plant with even greater respect.

Nightshade Family (*Solanaceae*). Though tobacco is the most valuable plant in the nightshade family, the potato is not far behind. Other important plants in the family are the tomato, pimiento, green pepper, and eggplant. Among the plants raised for their showy flowers are the petunia, Nicotiana, nightshade, cup of gold, and night-blooming jasmine. Many members of the family produce alkaloids, and some, such as belladonna, atropine, stramonium, hyoscyamine, and scopolamine, are important drugs. Nicotine extracted from tobacco is used as an insecticide. Mandrake, used as a remedy and pain killer from Biblical times, owes its properties to alkaloids. Cayenne pepper, red pepper, paprika, and chili powder are produced from various kinds of red peppers. Few Americans are acquainted with delicious naranjilla juice, squeezed from the orange-colored berries of *Solanum quitoense* which grows in the Andes. Though it is a favorite drink from Colombia to Peru it has never been processed for export and the plant will not grow well outside its native region.

It is noteworthy that our two most valuable vegetables in terms of dollar value of crop—potatoes and tomatoes—are both members of the nightshade family. Next to the cereal grains, the potato is our most important basic food. Though the United States produces around twenty-five billion pounds of potatoes a year, Russia, Poland, Germany, and France all far exceed this figure. Indeed, Europe produces about twenty times as many potatoes as America. Only bread is a more consistent menu item than potatoes. In addition to its prime importance as food, the potato is also the source of a variety of industrial products such as several kinds of alcohol, and of processed foods such as flour, starch, dextrins, and syrup.

Spurge Family (*Euphorbiaceae*). Although rubber can be made from the latex of many different plants, practically all of our natural rubber comes from *Hevea brasilensis,* the para rubber tree and the most important member of the spurge family. Though we are using more and more synthetic rubber, natural para rubber still is our most important source of this substance which is so essential in our modern economy. Many plants of the spurge family contain milky latex, and rubber is produced from several of them, particularly from Brazilian species of the genus *Manihot. Manihot exculenta* is a most important food plant in the tropical lowlands of South America, Africa, and the Far East, its root playing a role comparable to that of the potato in more temperate regions. It is variously called manioc, mandioca, cassava, and yuca, and it may be eaten boiled or whole or made into a meal similar to corn meal. It is also the source of tapioca flour and tapioca, the only form in which we use the plant. Candelilla wax is secured from *Euphorbia antisyphilitica,* which grows in the Big Bend country of Texas and northern Mexico.

Several members of the family are important sources of oils. Tung oil, long produced in China, is now extensively produced along our Gulf Coast and is most important as a drying oil for high quality paints and varnishes, though it is also used in making linoleum, oilcloth, printing ink, textile waterproofing, and electrical insulation. Castor oil, from another spurge, is a well-known medicine, but its many other uses are not so familiar. It is a superior lubricating oil and has recently been in great demand for use in jet engines. In addition, it is used in the same ways as tung oil, and in making nylon, soaps, imitation leather, plastics, flypaper, typewriter ink, and fabric finishes. Several other spurges, especially *Croton tiglium,* are also commercial sources of oils.

Several members of the family—including poinsettia, crown of thorns, and snow-on-the-mountain —are cultivated as ornamentals. Like the castor bean, the famous Mexican jumping bean is a member of the spurge family. Neither one is, of course, really a bean at all.

Rose Family (*Rosaceae*). This large and varied family is particularly outstanding for its fruits and ornamental plants. Roses themselves are perhaps

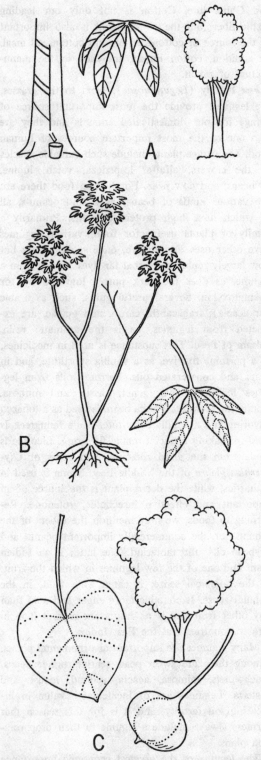

Fig. 2. Some valuable plants of the Spurge family. A–Rubber: gathering latex, leaf, tree. B–Cassava: plant and leaf. C–Tung: leaf, tree, fruit. (*From photo courtesy of David J. Rogers, New York Botanical Garden.*)

our leading flowers, and the family also provides such attractive ornamentals as the spiraeas, photina, pyracantha, mountain ash, hawthorns, and the flowering quinces, plums, peaches, and cherries. The list of rose family fruits sounds almost like an inventory of a fruit market: apples, peaches, plums, prunes, apricots, cherries, nectarines, quinces, strawberries, raspberries, blackberries, dewberries, youngberries, loganberries, and loquats. Apples are the most important fruit in this country and the most important tree fruit in the world as a whole, ranking second only to grapes. Almonds are also members of the family, and are not as out of place as they might seem. The almond seed, which we eat as a nut, is quite similar to the inedible peach seed in structure and appearance, and the surrounding dry and inedible fruit is structurally comparable with the fruit of the peach.

Madder Family (*Rubiaceae*). Unlike the rose family, the madder family owes its economic importance to only two genera: *Cinchona*, from which we secure quinine, and *Coffea*, (Fig. 3), the source

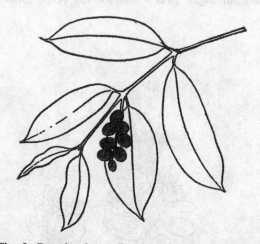

Fig. 3. Branch of a coffee tree showing a cluster of berries that contain the coffee beans.

of coffee. It is the billion-dollar value of coffee which is responsible for the inclusion of the family, even though quinine is really a more useful substance. It is still one of our most valuable drugs, even though it has been partly replaced by synthetics.

Yeast Family (*Saccharomycetaceae*). Up to this point all of our first families have been seed plants. Most of the plants which do not produce seeds, such as the ferns, mosses, algae, and fungi, are at present of considerably less direct economic value to man

than the seed plants. The fungi are, of course, extremely important economically in a negative way because of the decay they cause, and because of the diseases of man and his domestic plants and animals they produce, the losses running into billions of dollars a year. Certain fungi, however, are of very great positive economic value to man as sources of chemicals and in the processing of various foods. None are more important than the yeasts (Fig. 4), which have been said to be the first plants ever domesticated by man, though he certainly did not realize that he was dealing with a plant.

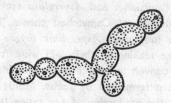

Fig. 4. Yeast cells, highly magnified. Yeast plants play the central role in two major industries: baking and brewing.

The great importance of the yeasts is due to the two products of their anaerobic respiration—carbon dioxide and alcohol. The first is important in the baking industry, the bubbles of carbon dioxide causing the dough to rise, while the second is the basis of the brewing industries. The combined values of the products of these two major industries, both dependent on yeast, easily put the yeast family into the billion-dollar class. In addition, yeasts are important sources of commercial vitamin supplies, and other complex chemicals are extracted from them commercially. It should be noted, too, that alcohol has a wide variety of industrial uses, entirely aside from its role in beverages. Some species of yeast are edible, having various pleasing flavors quite unlike those of ordinary brewer's or baker's yeast, and can be made into a wide variety of different types of dishes. As the pressure of the population on the food supply increases yeast may become more widely used as a food. Yeasts are already being produced in bulk in a few places such as the West Indies to supplement the limited food supply of the poorer people.

Blue Mold Family (*Aspergillaceae*). Next to the yeasts the several families of molds are the most useful fungi, and of the various molds the blue molds are the most valuable. The common family name has come from the bluish green color of the spores of some species, though other species in the family have spores of other colors such as black and pure green. Most of the molds growing on fruits belong to this family and the operator of a fruit market is likely to consider them only as a curse. However, the positive economic value of the family far outweighs the destructive activities of its members.

The oldest use of these molds was in processing of foods. Roquefort and blue cheeses owe their bluish green mottling and their characteristic flavor and odor to the presence of *Penicillium roquefortii*. *Penicillium camembertii* and a half-dozen other species of *Penicillium* and *Aspergillus* are involved in the production of Camembert cheese. The Japanese have long used *Aspergillus oryzae* and *A. flavus* in the making of their saki wine from rice. The molds convert the starch of the rice into sugar, which is then fermented by yeasts. Molds are also involved in the production of soy sauce from soybean seeds by the Chinese and Japanese.

Though food processing is an important activity of the blue molds, and is likely to become more so, their greatest importance is in the production of a variety of useful organic chemicals, particularly acids, enzymes, and antibiotics. *Aspergillus niger* and other members of the family, rather than citrus fruits, are the source of most of the commercial citric acid. Other acids produced by the blue molds include gluconic, gallic, and oxalic. Several members of the family, especially *A. clavatus*, produce alcohol in some quantity, though yeasts are the usual source. Takamine diastase, a starch-digesting enzyme produced by *Aspergillus*, is probably the best-known

enzyme secured from the blue molds, but proteolytic and pectic enzymes are also extracted commercially. By far the most important and valuable chemicals produced by fungi are the antibiotics, particularly penicillin, which is now largely secured from *Penicillium notatum* (Fig. 5), though other species also synthesize it. Most people are surprised when they learn that the monthly production of penicillin is measured in tons.

OTHER IMPORTANT FAMILIES

The selection of these ten first families has not been an easy matter. While any such list would include the first six or seven families, logical competitors might well be advanced for the remaining positions. Among these are the grape, citrus, palm, and lily families, as well as the banana family for both its fruits and its Manila hemp or abaca fiber. The beech family might well be considered for its beech and oak woods and cork, and the goosefoot family for sugar beets if not for red beets, Swiss

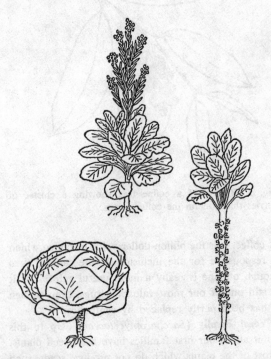

Fig. 6. The wild cabbage plant (top) is apparently the ancestor of our various cultivated cabbage relatives: cabbage (left) and Brussels sprouts (right).

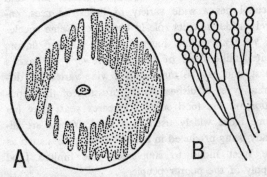

Fig. 5. A—Colony of *Penicillium notatum*, source of penicillin, in a culture dish shown inhibiting growth of bacteria inoculated over surface of culture medium. B—Small portion of a *Penicillium* plant under a microscope, showing the filamentous hyphae and green spores.

chard, mangel-wurzels, and spinach. The large sun-flower family provides many garden flowers, but is most important for lettuce. Several contending families contain a variety of useful plants. The gourd family provides us with squashes, pumpkins, cucumbers, cantaloupes, watermelons, and muskmelons. The mustard family is the source of table mustard and oils, as well as of many vegetables such as mustard greens, garden cress, water cress, radishes, horseradish, Chinese mustard, Chinese cabbage, rutabaga, and turnips. Even more important are cabbage (Fig. 6), Brussels sprouts, broccoli, cauliflower, and collards, all of them varieties of a single species (*Brassica oleracea*). The parsley family is

another major source of vegetables and herbs, including carrots, celery, parsley, parsnips, dill, anise, caraway, angelica, coriander, cumin, and fennel. The heath family provides cranberries, blueberries, huckleberries, wintergreen, rhododendrons, azaleas, mountain laurel, and trailing arbutus.

If you prefer to substitute any of these families for the last three or four in the list you are free to do so, though we shall stand by our original selections. In any event the statement that the ten first families exceed all the others in the value and importance of their products would still hold true because of the overwhelming importance of the six or seven leaders.

Chapter 2

THE KINDS OF PLANTS

No one botanist could, or would even want to, learn the name of each of the 350,000 species of plants, let alone become acquainted with their characteristic structural features, life cycles, and activities. Fortunately, it is possible to classify the many species of plants into a relatively small number of major related groups, each consisting of plants with certain common basic characteristics. This enables students of botany as well as botanists to gain some comprehension of the nature, scope, and diversity of the plant kingdom. In this chapter we shall survey some twenty major and quite distinct groups of plants, considering the common structural characteristics and processes, the general biological significance, and the economic significance of each of the groups. A few representative species of each group will be described and illustrated. We shall then consider the matter of plant classification and some of the problems involved in classifying plants.

To many people the word *plant* has a limited meaning, including farm and garden crops and herbaceous wild plants, but excluding such organisms as molds, bacteria, mushrooms, algae, and even trees. Similarly, many people restrict the term *animal* to mammals, excluding worms, protozoa, and other invertebrates and even vertebrates such as fishes, reptiles, and birds. Biologically, however, both terms have much broader meanings, all living organisms commonly being considered as either plants or animals. We are using the term *plant* in this broad sense here, and our survey of the plant kingdom will include a great variety of organisms ranging from bacteria to orchids. It is, however, quite difficult to decide whether some of the simpler and more primitive organisms should really be classed as plants or animals and so, as we shall see later in this chapter, some biologists prefer to divide living organisms into three or even four kingdoms, rather than the customary two.

The plant kingdom has at various times been divided into two to four principal groups on the basis of a variety of criteria. For our present purposes we shall classify plants as either **vascular** or **non-vascular,** a broad distinction somewhat comparable to the classification of animals as either vertebrates or invertebrates. Vascular plants are characterized by the presence of specialized vascular tissues that transport foods, water, and other substances from one part of the plant to another, and by the presence of true roots, stems, and leaves. Non-vascular plants lack true vascular tissues and true roots, stems, and leaves, and their structural patterns have a much greater range of diversity than those of the vascular plants. The vascular plants constitute a more advanced group from the standpoint of evolution and generally have a more complex structural organization than the non-vascular plants.

NON-VASCULAR PLANTS

Blue-Green Algae. The blue-green algae (Fig. 7) are among the simplest of all plants, consisting either of single cells or of a few cells arranged in variously shaped clusters or in threadlike **filaments,** commonly embedded in a gelatinous sheath. Their cells are smaller than those of most plants and are not differentiated into obvious cell structures such as nuclei and chloroplasts. However, blue-green algae contain **chlorophyll** and thus can make food by photosynthesis. In addition they characteristically contain a blue pigment (**phycocyanin**) that in combination with the chlorophyll imparts the dark blue-green color distinguishing the group. Some blue-green algae also contain a red pigment (**phycoerythrin**), and in a few species this is concentrated enough to give the plants a red color. Blue-green algae lack typical sexual reproduction, being propagated by cell division or the fragmentation of filaments into two or more parts. Some blue-green algae are able to fix atmospheric nitrogen (convert the N_2 to nitrogen compounds, especially NH_3), apparently a capacity not present in other photosynthetic plants. Thus they are the most self-sufficient of all organisms.

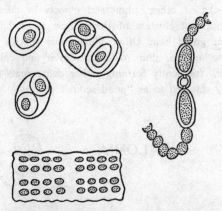

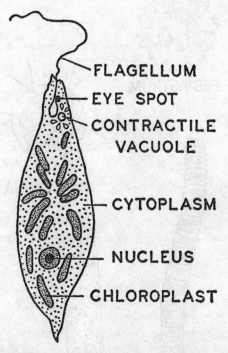

Fig. 8. *Euglena*. The cell may also contract until it is almost spherical.

Fig. 7. Blue-green algae. UPPER LEFT: *Gleocapsa*. LOWER LEFT: *Merismopedia*. RIGHT: *Anaboena*.

There are about fourteen hundred species of blue-green algae, most of them living on or below the surface of moist soil, on tree trunks or rock surfaces, or in stagnant pools of water. Several species live in the water of hot springs at a temperature (175° F.) that would kill most organisms. Blue-green algae may serve as food for various animals, and those species that fix nitrogen contribute to soil fertility, but the general biological significance of the blue-green algae is relatively minor when compared with many other groups of plants. At present the blue-green algae have essentially no economic value, but they cause trouble by growing in reservoirs and imparting a disagreeable odor and an unpleasant "fishy" taste to the drinking water.

The Euglenoids. This group includes only about 235 species, characteristically unicellular organisms that swim by means of whiplike **flagella** attached to the front end of the cell. Euglena (Fig. 8) is the best-known representative of the group. It lacks a cell wall, has an anterior gullet with a red eye-spot that is apparently sensitive to light, and has a distinct nucleus and chloroplasts. The chloroplasts do not contain pigments other than the chlorophylls and carotenoids present in higher plants and so have a

grass-green color. Because of the absence of a cell wall euglena can change shape rather freely, varying from an elongated spindle, through something of a pear shape, to an almost spherical form. Some of the other species, however, have cell walls, while still other species lack chlorophyll and are nourished by ingesting food through the gullet in the manner of protozoa. Indeed, some biologists regard the euglenoids as being protozoa rather than plants. It is certainly difficult to decide whether they are plantlike animals or animal-like plants. The euglenoids have no direct economic importance, but they are quite abundant in ponds, lakes, and streams and provide an important source of food for fish and other aquatic animals.

Green Algae. The green algae constitute a large group (almost six thousand species) of plants widely distributed in pools, ponds, lakes, streams, soil, and on the surfaces of rocks and tree trunks. Though most species live in fresh water, there are a considerable number of marine species. The green algae have a wide range of structural patterns, including single-celled species, clusters of cells constituting colonies, and filaments (Fig. 9). In some species the filaments are branched. Sea lettuce, a marine-green algae, represents still another structural pattern, the

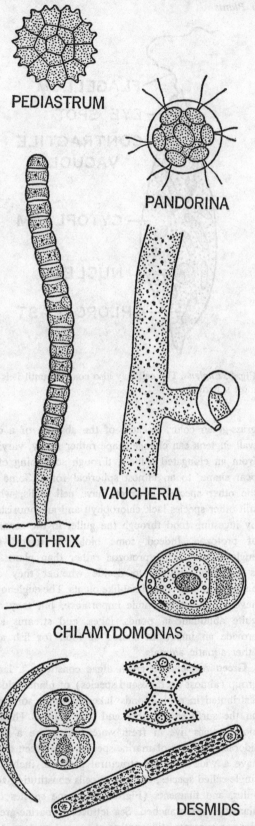

PEDIASTRUM

PANDORINA

VAUCHERIA

ULOTHRIX

CHLAMYDOMONAS

DESMIDS

Fig. 9. Several species of fresh-water algae.

main part of the plant consisting of a sheet of cells only two cells thick but quite long and broad, one species of sea lettuce being over three feet long (Fig. 10). Most green algae, however, like the blue-green algae and euglenoids, are microscopic, though the masses of plants floating on the water or attached to submerged objects are readily visible. Some filamentous species have specialized **holdfast** cells at the lower end whereby they are attached to rocks or other submerged objects in lakes or streams, and clusters of these algae somewhat resemble green hair. Other filamentous species float on the surface, along with unicellular and colonial species, frequently forming rather dense mats commonly referred to as "pond scum."

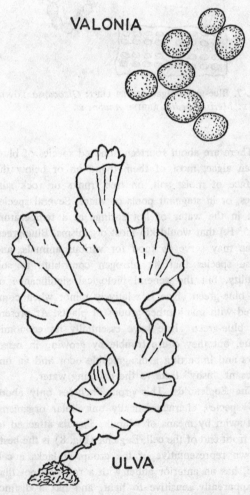

VALONIA

ULVA

Fig. 10. Two rather large green algae that live in the oceans. The *Valonia* cells are an inch or so in diameter, each cell having numerous nuclei. The *Ulva* (sea lettuce) may be a foot or more long.

All green algae have organized nuclei and chloroplasts, the latter being of varied and often rather striking shapes (Fig. 9). The chloroplasts contain chlorophylls and carotenoids, like the euglenoids and higher plants. The cell walls are composed basically of cellulose, as in the higher plants. In addition to reproducing asexually by cell division or motile spores, practically all green algae also reproduce sexually, in contrast with the blue-green algae.

The green algae are of great biological significance since they provide a major portion of the food of fresh-water animals and play an essential role in maintaining the supply of oxygen used by aquatic organisms. At the present time they are of little direct economic importance, but their future economic value may be very great. For some time there has been extensive experimentation with unicellular algae, principally *Chlorella,* directed toward increasing growth rate, photosynthesis, and yield, and toward designing suitable systems in which algae can be cultured in quantity as a source of human or animal food. Algae may be an important human food source in the future, as the population of the earth keeps increasing, but a more immediate possible use is in providing both food and oxygen for the occupants of manned satellites.

Diatoms and Their Relatives. The yellow-green algae are mostly unicellular plants, though some species are colonial or multicellular. There are three classes: the true yellow-green algae, the golden brown algae, and the diatoms. Altogether there are about fifty-seven hundred species, over five thousand of them being diatoms. Most species live in fresh water, though some are found on moist soil, tree trunks, or rocks. Some species are marine, the diatoms in particular constituting an important part of the flora of the oceans. The cell walls are usually composed of overlapping halves, like the bottom and lid of a box, and particularly in the diatoms are impregnated with silica. In addition to chlorophyll, the chloroplasts contain substantial quantities of yellow or brown pigments that impart colors characteristic of this group. Though diatoms are not flagellated, many of the species in the other two classes are. However, some diatoms can move about in a jerky series of glides.

The different species of the diatoms have a wide range of attractive geometrical shapes (Fig. 11) and there are generally intricate and characteristic markings on their siliceous walls. The arrangement of

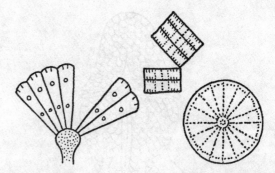

Fig. 11. Several species of diatoms, highly magnified.

these is so precise and regular that manufacturers of microscope lenses use diatoms to determine whether the lenses are free from distortion. Diatoms have great biological importance as one of the major sources of food for both marine and fresh-water animals; they are probably the most important food producers in the oceans. After marine diatoms die their walls fall to the bottom of the ocean and since the silicon portion does not decay, thick deposits of **diatomaceous earth** may eventually accumulate. Such deposits are found in land areas in various parts of the world that were ocean bottom in past geologic ages and these deposits are quarried for commercial use. Diatomaceous earth is extensively used in the manufacture of fine abrasives and polishes, as insulation material, as an absorbent carrier in industrial explosives, and as a filtering agent in oil and sugar refining and other industrial processes.

The Dinoflagellates and Relatives. Most of the species in this group are unicellular, flagellated, marine organisms. There are about a thousand species, most of them dinoflagellates (Fig. 12), and all of these having the three characteristics mentioned. Some species have cell walls while others do not, the walls being basically cellulose. Reproduction is commonly asexual by cell division. The chloroplasts contain yellow or brown carotenoids as well as chlorophyll. The walls of dinoflagellates are composed of a definite number of sections or plates joined together, and the cells are also characterized by a groove around them with two flagella attached in or near the groove.

Next to the diatoms, the dinoflagellates are probably the most important food source of marine animals. One dinoflagellate that produces a red pigment periodically increases greatly in number, resulting in the "red tide" that kills vast numbers of fish.

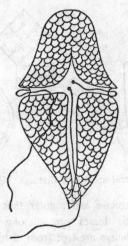

Fig. 12. A dinoflagellate.

The Brown Algae. The brown algae (Fig. 13) are almost all marine, most species growing attached to rocks along the shores of the cooler seas though some species grow at depths of almost three hundred feet. A few species float freely in the open ocean, the best known being sargassum, the seaweed that is so abundant in the famous Sargasso Sea of the South Atlantic Ocean. About nine hundred species of brown algae are known, ranging from microscopic ones to large, leathery kelps that may attain lengths of two hundred feet or more. Most of the larger brown algae are extremely tough, leathery, and rather coarse plants with branched, rootlike **holdfasts,** stemlike **stipes,** leaflike or strap-shaped **blades,** and air-filled **floats.** Each species has a characteristic arrangement of most or all of these parts, a great variety of forms being found. One of the more striking is the sea palm of the west coast, which has a thick stipe and a crown of blades making it resemble a palm tree in general shape. Most species are olive-green to brown in color because the chloroplasts contain a brown carotenoid pigment (fucoxanthin) in addition to the more common yellow carotenoids and chlorophylls *a* and *c*. All brown algae are multicellular, some cell differentiation occurring in the larger species. Some species have specialized food-conducting tissues. The cells are similar to those of higher plants, and have vacuoles and a single nucleus each. The cell walls are basically cellulose, but they also contain abundant gelatinous substances. All species have sexual reproduction, but some also can reproduce asexually by spores or fragmentation.

As a food source for marine animals the brown

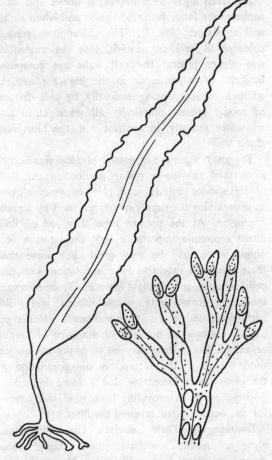

Fig. 13. Marine brown algae. LEFT: *Laminaria*. RIGHT: *Fucus*.

algae are much less important than the diatoms and dinoflagellates. Some brown algae are used as human food in various parts of Asia. Algin extracted from brown algae is used as a stabilizing agent in ice cream and other foods, and as jelling, thickening, and emulsifying agents in various foods, medicines, tooth pastes, and lotions. Brown algae contain high concentrations of some elements such as iodine and potassium, and have been used as sources of these elements.

The Red Algae. Along with the brown algae, the red algae (Fig. 14) include most of the plants commonly called "seaweed." The characteristic red or pink color of the red algae results from the presence of a red pigment (phycoerythrin) in the chloroplasts, in addition to chlorophylls *a* and *d*, carotenoids, and sometimes the blue phycocyanin. The phycoerythrin is usually concentrated enough to provide the dominant color. Like the brown

the preparation of jellylike culture media for bacteria and fungi in biological laboratories, as a laxative, in sizing textiles, and for many of the same uses as are made of algin.

Bacteria. Though people commonly think of bacteria as animals, biologists are in general agreement that they belong in the plant kingdom rather than in the animal kingdom. Bacteria are typically unicellular, though some species have chains or clusters of cells. Some bacteria are spherical (**coccus** forms), others are rod-shaped cells (**bacillus** forms), while a few species have spiral cells (**spirillum** forms) (Fig. 15). Bacterial cells are much smaller than

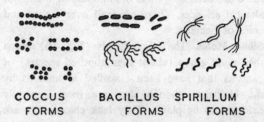

COCCUS FORMS   BACILLUS FORMS   SPIRILLUM FORMS

Fig. 15. Several species of bacteria showing the three basic cell shapes: coccus, bacillus, and spirillum.

Fig. 14. Two species of red algae, both marine. The red algae are generally delicate in texture as well as in form and range up to a foot or two in length. TOP: *Grinnelia.* BOTTOM: *Gelidium,* a principal source of agar.

algae, most red algae are multicellular, but they are mostly somewhat smaller and more delicate. The range of form from species to species is great, including flat or wavy sheets, ribbons, delicate cylinders, and highly divided featherlike structures. Some twenty-five hundred species are known, practically all marine. They generally grow at the bottom of the ocean, often as deep as six hundred feet, though some are found on rocky shores. Some species become calcified and contribute to the formation of coral reefs. The cell walls are composed of cellulose and are commonly impregnated with a variety of jellylike or mucilaginous substances. Most species reproduce sexually.

Red algae are consumed by fish, and even land animals may graze on those growing along the shore lines. These plants have a limited use as human food: Irish moss and dulse (or sea kale) are red algae. However, the red algae are most important economically as the source of **agar,** widely used in

those of most other organisms, an ordinary parenchyma cell from the stem of a plant having room in it for two thousand or more bacteria. Bacterial cells are relatively simple in structure, lacking a well-defined nucleus or other cell structures visible under an ordinary microscope. The cells have walls containing cellulose in some species. The outer layer of the wall may be a thin slimy material called the **capsule,** and some species are embedded in an extensive gelatinous mass. Some species of bacteria are motile, being propelled by a whiplike movement of their threadlike flagella. Bacteria commonly reproduce by cell division, one cell dividing into two in as short a time as twenty minutes under favorable conditions. If this rate were maintained for twenty-four hours there would be almost five billion billion (5×10^{21}) offspring weighing about four million pounds. Such rapid reproduction is never actually attained because of the exhaustion of available food and accumulation of toxic waste products. At least some species of bacteria also have a primitive type of sexual reproduction.

Bacteria are found almost everywhere on earth—on most objects, in the air, in the soil and bodies of water, and in the digestive tracts of animals (where some species aid in digestion). Many bacteria are **parasitic,** invading the tissues of plants or animals and frequently causing diseases, but most bacterial species are **saprophytic,** securing their food from non-living organic matter. A few species are **autotrophic,** making their own food either by photosynthesis or a similar process (chemosynthesis) not requiring light energy.

Bacteria have great economic importance as causes of disease, food spoilage, and decay, in the commercial production of various chemicals; in the production of cheeses, vinegar, and other foods, and in the maintenance of soil fertility. Because of the economic and biological importance of bacteria the science of bacteriology has become one of the major biological sciences. There are well over a thousand species of bacteria.

Slime Molds. The slime molds (Fig. 16) are a rather small group (about four hundred species) of organisms that have been classified as animals by some biologists, though they are more generally considered to be plants. They lack chlorophyll, are

consist of a slimy and rather sticky mass of protoplasm (the **plasmodium**) without cell walls, that may cover an area of several square inches (Fig. 17). Each plasmodium contains many nuclei. Some-

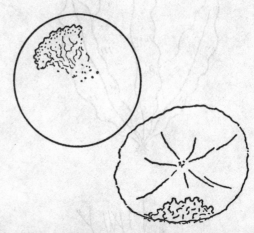

Fig. 17. Plasmodia of slime molds. RIGHT: A slime mold in its natural habitat, moving slowly over the cut end of a log. LEFT: Plasmodium of *Physarum* cultured in a Petri dish. Later these plasmodia develop into sporangia similar to those shown in Fig. 16.

LAMPRODERMA

STEMONITIS

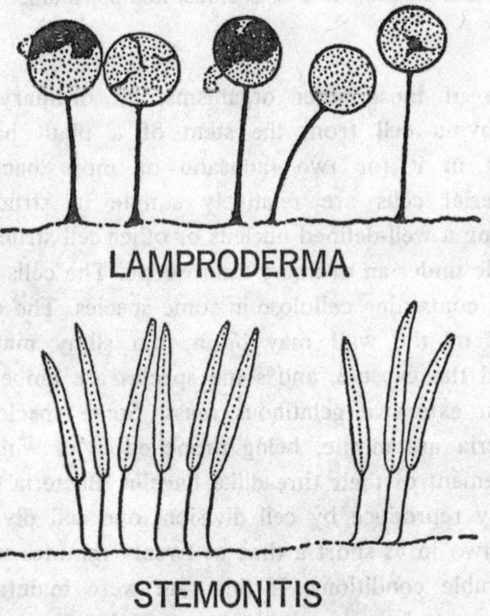

Fig. 16. Sporangia of two species of slime molds. The sporangia range between a quarter and a half inch in height and may be brightly colored in some species.

mostly saprophytic, and are found principally on organic substances in soils, on dead leaves, and on decaying wood and bark. They are most evident during warm, rainy periods. During the vegetative (non-reproductive) stage of their life cycle they

times the plasmodium has a branched appearance, while in one type it assumes an elongated sluglike form. The plasmodia move by the flowing of the protoplasm and may engulf particles of food in vacuoles, as in the case of amoebae. Some species of slime molds occur as individual cells very similar to amoebae in size and appearance at one stage of their life cycle, and then hundreds of these individual cells come together and form a single plasmodium (Fig. 17). The vegetative stages of the slime molds have characteristics that qualify them for classification as animals, but the reproductive structures (the spore-bearing **sporangia**) are definitely plantlike structures. The sporangia of most slime molds are less than 1 centimeter high, though a few have sporangia as large as 7.5 cm. The sporangia have a wide variety of shapes and colors characteristic of the species (Fig. 16). Though many species are white, others are violet, purple, orange, yellow, or brown.

A few species of slime molds are parasitic on fungi or higher plants. The clubroot disease of cabbage and the powdery-scab disease of potatoes and tomatoes are caused by slime molds and at times result in considerable crop reductions. Otherwise the slime molds are of little economic importance, nor do they play an important role in the general biological scheme of things. However, slime molds have been of great interest to biologists because of their unique and relatively simple structure. They are excellent subjects for the study of development and differentiation and for the study and chemical analysis of substantial quantities of protoplasm unencumbered by cell walls.

The True Fungi. The true fungi constitute a large and diverse group of plants numbering over seventy-five thousand known species. The name of the group distinguishes it from "fungi" in the broader sense, a term often considered to include the bacteria and slime molds as well as the true fungi. The true fungi (which we shall from now on call simply "fungi") all lack chlorophyll and are either parasites or saprophytes. Most species of fungi live on land as parasites on other organisms or on organic matter, wherever it occurs, as saprophytes, but many species live in bodies of both fresh and salt water on organisms or their products. Most fungi are composed of filaments called **hyphae,** and the mass of hyphae making up a fungus are referred to collectively as the **mycelium.** In such fungi as mushrooms and bracket fungi the hyphae are matted together tightly and form fleshy or woody structures of considerable size and substance, while in other

fungi, such as bread molds, the individual hyphae are more distinct and form a cottony fluff. A few fungi like the yeasts (Fig. 4) commonly occur as single cells or short filaments, lacking true mycelia. All fungi reproduce asexually by one kind of spore or another, the water molds commonly having flagellated spores. Most fungi also reproduce sexually, the sexual structures being quite varied and the processes of sexual reproduction commonly being highly modified and frequently very complex.

There are three quite distinct classes of fungi: the **phycomycetes** (or algalike fungi), the **ascomycetes** (or sac fungi), and the **basidiomycetes** (or club fungi). The phycomycetes (Fig. 18) are somewhat simpler in structure than the other two classes and their mycelia are often very limited in extent or at most loose, cottony masses, rather than compact masses as in many members of the other two classes. The hyphae of the phycomycetes are **coenocytic,** that is, they have no cross walls except in reproductive structures although many nuclei are present in a hypha. Some species are unicellular. Most species have both sexual and asexual reproduction, the asexual spores of species that live in the water generally being flagellated, motile **zoospores,** while the terrestrial species have air-borne spores. The parasitic species of phycomycetes live on other

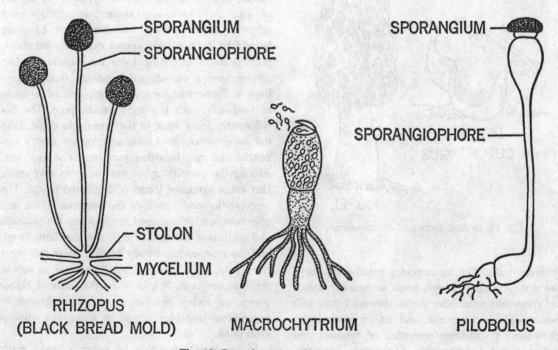

Fig. 18. Several species of phycomycetes.

fungi, algae, higher plants, or animals, while the saprophytic species live mostly in water, moist soil, dung, dead organisms, and on foodstuffs. Among the better-known phycomycetes are water molds, such as those that parasitize fish or grow on dead insects in the water, the downy mildews that are parasites of higher plants, and the common black bread mold (*Rhizopus nigricans*).

In contrast with the phycomycetes, the hyphae of the ascomycetes (Fig. 19) are divided into cells

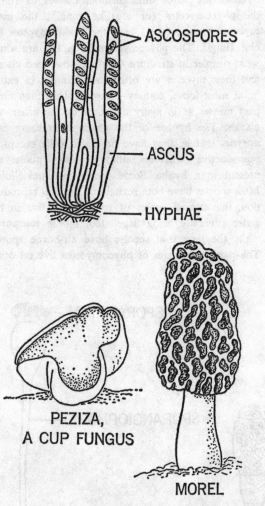

Fig. 19. Several species of ascomycetes.

by cross walls. The ascomycetes constitute a large and diverse group, ranging from structurally simple and frequently unicellular yeasts, through forms with loosely organized mycelia, and on to species with spore-bearing structures composed of compact and well-organized fleshy mycelia. However, all ascomy-

cetes have a characteristic reproductive structure that sets them apart sharply from other fungi: the saclike **ascus** that bears four or eight **ascospores** internally (Fig. 19). These ascospores are a stage in the sexual life cycle. Many ascomycetes also produce asexual spores. Among the parasitic ascomycetes are yeasts and related species that cause serious diseases of the skin, nervous system, ears, lungs, or other organs of man and animals, as well as the powdery and sooty mildews, black rot, chestnut blight, and Dutch elm disease fungi that parasitize higher plants. Among the other ascomycetes parasitic on higher plants is the ergot of rye, wheat, and other grains. The grains of plants parasitized by ergot become filled with ergot hyphae that contain a very toxic substance that poisons or even kills people who use flour made from grains containing ergot. However, ergot is a source of an important drug that checks hemorrhages when used in proper doses. Among the saprophytic ascomycetes are the yeasts so important in brewing and baking and as a source of vitamins, the orange bread mold that has been so useful in studies of heredity, the blue and green molds that cause the decay of fruits and include *Penicillium* (Fig. 5), the source of penicillin, the cup fungi, and several edible mushrooms such as the morels and truffles.

The basidiomycetes (Fig. 20) are similar to the ascomycetes in having cellular hyphae, but instead of asci they have a characteristic spore-bearing structure known as the **basidium.** Each basidium bears four (rarely two) spores externally on short, pointed stalks extending from the basidium. Most species have a club-shaped basidium, though some have a filamentous basidium composed of a chain of four cells, each bearing a basidiospore. The basidiospores are a stage in the sexual life cycle. Like the ascomycetes, the basidiomycetes are mostly terrestrial and may be either parasites or saprophytes. Among the parasitic types are the rusts and smuts that cause extensive losses of cultivated plants. The saprophytic species include the common edible and poisonous umbrella-shaped mushrooms, the puffballs and earthstars, the tooth fungi, the stinkhorn fungi, and the more-or-less woody bracket fungi and other pore fungi. Some species of basidiomycetes, as well as some ascomycetes, live in or on the roots of higher plants, producing root-fungus complexes known as mycorhizae that play a role in water and mineral absorption.

It has been impossible to assign a good many

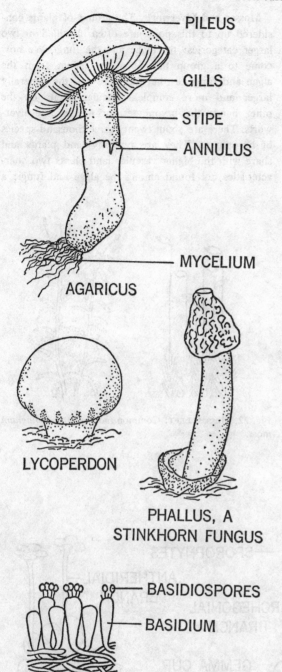

AGARICUS

- PILEUS
- GILLS
- STIPE
- ANNULUS
- MYCELIUM

LYCOPERDON

PHALLUS, A
STINKHORN FUNGUS

- BASIDIOSPORES
- BASIDIUM

Fig. 20. Several species of basidiomycetes.

species of fungi to one of the three classes just described, principally because their complete sexual life cycle is unknown. These are grouped together

as the **imperfect fungi,** pending more adequate classification. Most of them probably are really ascomycetes, though no asci have been found. The imperfect fungi include many parasitic species that cause plant diseases and also species that cause such diseases of man as ringworm, athlete's foot, thrush, sprue, and a variety of serious lung infections.

Lichens (Fig. 21) are composite organisms composed of unicellular algae growing in among the hyphae of a fungus, the relationship being parasitic (but not pathogenic) or perhaps symbiotic. Note

FOLIOSE LICHEN

FRUTICOSE LICHEN

Fig. 21. Two kinds of lichens.

that the form and structure of lichens are mainly determined by the fungus rather than the alga. In general, the lichen fungi are known only in their lichen occurrence, while the lichen algae may also occur separately. Despite the fact that lichens live and reproduce as a unit, they are considered as a complex of two organisms rather than a single organism. For this reason they are now classified with the fungi rather than as a separate group, as had been done in the past. The fungal part of most lichens is an ascomycete, though a few are basidiomycetes. Lichens are common on tree trunks and rocks, generally having a flattened crusty, flaky, or leaflike appearance, but there are also some larger and highly branched lichens such as the reindeer moss. This and other large lichens are an important food source for arctic animals. The manna of the Bible is believed to have been a lichen. Lichens have been used commercially as sources of dyes, tannins, drugs, cosmetics, and litmus, the indicator used in laboratories.

The fungi in general are a most important group economically, not only because of the plant and animal diseases they cause, but also because they are the source of many important items of commerce such as penicillin and other antibiotics, alcohol, citric acid and many other chemicals, and the mushrooms used as food. Fungi also play an essential role in the making of many different cheeses, such as Roquefort. The decay caused by fungi plays a useful role in nature in disposing of the remains of plants and animals, but is less welcome when decay fungi attack foods, lumber, fabrics, paper, leather, or other items of commerce valuable to man.

Mosses and Liverworts. The groups of plants considered up to this point are often classified in two larger categories: the algae and the fungi. We now come to a group that is quite distinct from the algae and fungi on the one hand and the generally larger and more complex vascular plants on the other hand: the bryophytes, or mosses and liverworts. There are about twenty-three thousand species of bryophytes. They are typically land plants and share with the higher vascular land plants two characteristics not found among the algae and fungi: a

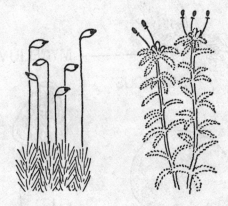

Fig. 22. Mosses. LEFT: Common moss. RIGHT: *Sphagnum* moss.

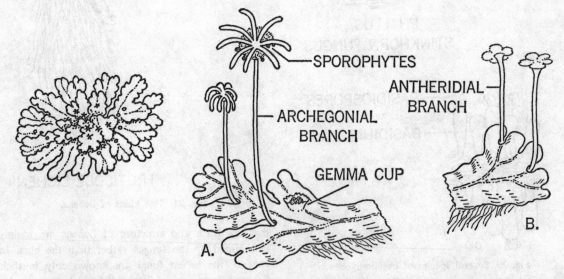

Fig. 23. Liverworts. LEFT: A liverwort growing on a moist stream bank, showing the flat, leaflike, much-branched thallus. RIGHT: Portion of *Marchantia* showing female (A) and male (B) plants with reproductive structures (antheridial and archegonial branches). The sporophytes are small structures borne on the archegonial branch.

multicellular embryo plant and multicellular sex organs bearing eggs and sperm. Unlike the vascular plants, however, bryophytes do not have true roots, stems, or leaves or special tissues that transport water and food (vascular tissue). The stemlike and leaflike structures of mosses are not true stems or leaves. Mosses (Fig. 22) usually grow in dense mats in shaded, moist places and thrive particularly in acid soils. They are generally found on the surface of the soil or on rocks. Liverworts (Fig. 23) are usually found in even moister locations than mosses, many species growing along rocky stream banks where they are splashed by water, or even under water. Bryophytes are found in temperate regions, but are most abundant in tropical rain forests. The liverworts are mostly rather flat leaflike structures quite different in appearance from the erect leafy stalks of the mosses.

Liverworts and mosses contain chlorophyll and make their food by photosynthesis, though some species also absorb decaying organic matter. The bryophytes are of little economic value compared with most other plant groups, though they do contribute substantially to the prevention of soil erosion. The most useful moss is the large peat moss, sphagnum. It is used in packing such items as china or glassware and as a moist packing for cut flowers or nursery stock, in making absorbent surgical dressings, and in increasing the organic matter of garden soils. Peat, the partially decayed remains of sphagnum moss and other plants, is widely used as a fuel in Ireland, Scotland, and some other countries.

The bryophytes, like all vascular plants and most algae and fungi, have a reproductive life cycle with two distinct phases or stages: a **gametophyte** phase that reproduces sexually by means of gametes (eggs and sperm) and a **sporophyte** phase that reproduces by means of a special type of spore (**meiospores).** In bryophytes, as in most green algae, the gametophyte phase is the dominant one, the "leafy" moss plant being a gametophyte. The moss sporophyte consists only of a stalked sporangium that may be seen growing from the tips of female gametophyte plants (Fig. 22). In the vascular plants that we are about to consider, however, the sporophyte phase is always the dominant and conspicuous one, and in the higher vascular plants the gametophytes are reduced to microscopic structures. All of the vascular plants we shall describe are the sporophyte phases.

VASCULAR PLANTS

About 250,000 of the 350,000 or so species of plants so far identified are vascular plants. The dominant and conspicuous plants of our forests, grasslands, and deserts are all vascular plants, as are all plants commonly cultivated in farms and gardens. The dominating position of the vascular plants in the land vegetation of the earth is related to the presence of vascular tissue through which water, minerals, and food can be transported rapidly over considerable distances; the development of roots as effective anchoring and absorbing organs;

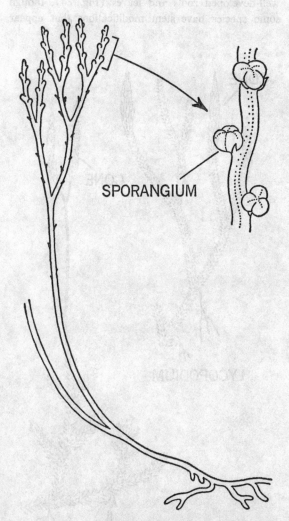

Fig. 24. *Psilotum,* probably the most primitive of all living vascular plants. A few of the many branches of a plant are shown. There are no leaves, but note the sporangia.

and the waterproofing of tissues in contact with the air. Only vascular plants have true roots, stems, and leaves; flowers, fruits, and seeds are limited to the more advanced vascular plants.

The Psilophytes. The psilophytes are an extremely small group of vascular plants, consisting of only two living genera with a total of three species, but around 360 million years ago the psilophytes were much more abundant. The living species are of little economic or ecological importance, but they are of great interest to botanists as being the most primitive vascular plants, probably similar to plants that provided an evolutionary connecting link between the more advanced green algae and higher vascular plants. The psilophytes have stems but lack well-developed roots and leaves (Fig. 24), though some species have stem modifications that appear

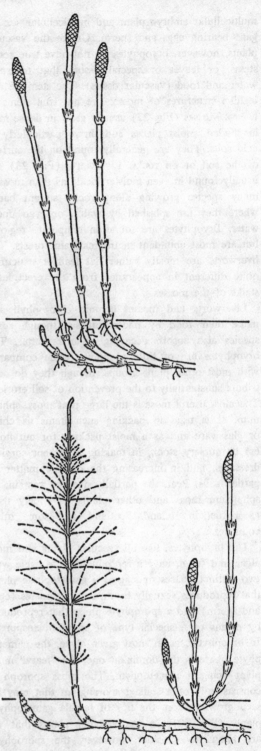

Fig. 26. Two species of *Equisetum*. BOTTOM: A species with cones borne on reproductive branches, that appear earlier than the green and branched vegetative branches. TOP: A species bearing its cones at the ends of the vegetative branches.

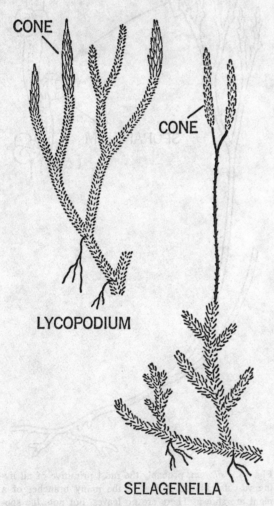

CONE

CONE

LYCOPODIUM

SELAGENELLA

Fig. 25. Two species of club mosses. Note the cones.

to be the forerunners of true leaves and roots. Thus, the stem appears to be the basic vegetative organ of vascular plants, with leaves and roots as modified derivatives of the stem. The ancient psilophytes are presumed to have given rise to at least three main evolutionary lines, one leading to the club mosses, one to the horsetails, and the other to the ferns and seed plants.

The Club Mosses. The nine hundred species of living club mosses (Fig. 25) are all low-growing herbaceous plants less than a foot tall, though some species have trailing stems many feet long. However, about three hundred million years ago there were many large tree species, with trunks up to 135 feet high and six feet in diameter. These ancient giant club mosses contributed substantially to our coal deposits. Club mosses have numerous small, simple leaves arranged spirally on the stems, distinct but rather poorly developed roots, and sporangia borne on leaves. In some species the spore-bearing leaves are at the tips of the stems and may lack chlorophyll, thus forming recognizable cones, while in other species the ordinary foliage leaves along the stem are spore-producing. Though club mosses are of world-wide distribution, they are most abundant in tropical and subtropical regions and commonly grow in forests. The club mosses have little economic importance, though the so-called running cedar or ground pine used for Christmas decorations in some parts of the country is a club moss.

The Horsetails. The horsetails (Fig. 26) constitute a very small group of plants, with only one living genus (*Equisetum*) and about twenty-five species. Like the club mosses, the horsetails were much more abundant about three hundred million years ago than they are now, and there were tree species growing up to ninety feet tall. These ancient horsetails made important contributions to the coal beds. The horsetails have ribbed and jointed stems impregnated with silica and are generally quite stiff and abrasive in texture. Some species have numerous unbranched vertical stems rising from the rhizomes, while others have stems that are highly branched. The bushy appearance of the latter type gave rise to the name "horsetail." The stems are green and carry on most of the photosynthesis, the leaves being small, papery, and commonly lacking chlorophyll. These small leaves are attached to the stem in whorls at the jointed nodes, one whorl being separated from the next by a rather long internode. The spore-bearing leaves are borne in short cones at the tips of the stems. The branched types usually have special reproductive stems that bear the cones. These reproductive stems generally appear early in the spring before the vegetative branches and are usually yellowish rather than green. The horsetails

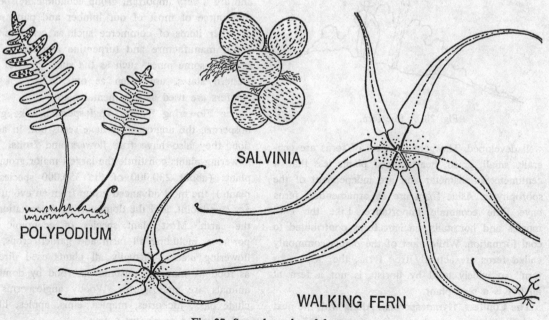

SALVINIA

POLYPODIUM

WALKING FERN

Fig 27. Several species of ferns.

are of practically no economic value, though pioneer women frequently used them for scouring pots and pans. This gave rise to another common name, "scouring rushes."

The Ferns. The ferns (Fig. 27) constitute a relatively large (almost ten thousand species) and well-known group of plants. Though most abundant in the tropics, they are of world-wide distribution and are commonly found in moist, shady places. Most ferns are terrestrial, but a few species grow in fresh water. In contrast with the club mosses and horsetails most species of ferns have large leaves, the spores being borne on the underside of the leaves. Some species have specialized spore-bearing leaves quite different in appearance from the foliage leaves, but the spore-bearing leaves are not arranged in cones. Some species of tropical ferns are trees (Fig. 28) with stems up to seventy-five feet high, but practically all of the ferns of temperate climates have underground stems and only the leaves appear above ground. Fern roots are generally quite

Fig. 28. A tree fern.

well developed. The gametophytes of ferns are generally small, green, heart-shaped plants less than a centimeter in diameter and are independent of the sporophytes. Aside from use as ornamentals, ferns have little economic importance. Like the club mosses and horsetails, ancient ferns contributed to coal formation. While most of the plants commonly called ferns are actually true ferns, the "asparagus fern" so widely used by florists is not a fern at all but is a seed plant.

The Conifers (Gymnosperms). Along with the next and final group of plants, the gymnosperms have

seeds, a characteristic setting them apart from the other vascular plants. The seeds of most gymnosperms are borne on the upper surface of the rather woody scales of the cones and are not enclosed in fruits. The pollen of gymnosperms is borne on separate and usually smaller cones than those that bear the ovules that develop into seeds. Gymnosperms are large, woody plants, most species being trees. There are about 725 species of gymnosperms in 63 genera, including many common and well-known ones such as pine, spruce, fir, hemlock, cedar, cypress, redwood, and arborvitae. Rather different in appearance and characteristics are a few other gymnosperms such as the cycads of tropical and subtropical regions and the ginkgo tree, a native of China widely planted in this country as a street tree. Unlike most conifers, which have needle-shaped or small leaves, the cycads have large, divided leaves while ginkgo has broad leaves.

Another rather unusual gymnosperm is the shrubby *Ephedra* of arid regions, the source of the drug ephedrine used in treating colds. The Gnetums are tropical vines, shrubs, or small trees with broad leaves. The most unusual gymnosperm, and indeed one of the most unusual of all plants, is *Welwitschia* of southwestern Africa. This plant has large, tuberous, woody undergound stems, a long root, and only two long strap-shaped leaves that lie on the ground.

The gymnosperms are of world-wide distribution and are a very important group economically, being the source of most of our lumber and paper and of other items of commerce such as cellulose for rayon manufacture and turpentine and resin. The seeds of some pines, such as the sugar pine of the western states, are eaten as nuts. Many of the conifers are used as ornamental plants.

The Flowering Plants (Angiosperms). Like gymnosperms, the angiosperms have seeds but, in addition, they also have true **flowers** and **fruits**. The flowering plants constitute the largest major group of plants (about 230,000 of the 350,000 species of plants), the most advanced group from an evolutionary standpoint, and the dominant land vegetation of the earth. Most plant species of economic importance, including all farm and garden crops, are flowering plants. Virtually all plants used directly as food by man or consumed as food by domestic animals are angiosperms. Woody angiosperms include oaks, hickories, maples, elms, apples, lilacs, and palms to name just a few.

Most angiosperms have broad leaves, in contrast with most gymnosperms. Though the cones of gymnosperms can technically be considered as flowers, the typical sepals, petals, stamens, and pistils of flowers are restricted to angiosperms. However, as will be noted in Chapter 4, not all angiosperm flowers have all these structures.

Two rather distinct types of angiosperms can be identified: the **monocotyledons** (or **monocots**) and the **dicotyledons** (or **dicots**). As their names imply, the basic distinguishing characteristic is the number of seed-leaves or cotyledons of the embryos and young plants. In addition, however, there are other characteristics that are generally quite different in one than the other. Monocots typically have their flower parts in threes or sixes, while dicots generally have theirs in fours or fives or some multiple of these. Monocots generally have parallel-veined leaves while dicots usually have netted-veined leaves. There are also differences in anatomy that will be considered later. Among the common monocots are members of the grass, lily, orchid, sedge, and palm families. Most monocots are relatively small and frequently succulent herbaceous plants, but there are a few woody monocots including palm trees, Joshua trees of the arid southwest, and bamboo. Most woody angiosperms are dicots, including the majority of our common trees and shrubs. There are also many herbaceous dicots, including members of the potato, carrot, mint, morning-glory, pea, mallow, spurge, and sunflower families.

The vast assemblage of angiosperms includes a number of quite strange and unusual plants. These include the many species of cactus, the tiny duckweeds that float on ponds and lakes, the Spanish moss, which is a member of the pineapple family but grows draped over tree branches, and the white Indian pipe that lacks chlorophyll and lives as a parasite on a fungus.

CLASSIFICATION OF PLANTS

Whenever one deals with groups of things as large as the 350,000 or more species of plants some system of classification is essential. If a library of 350,000 volumes had the books arranged on the shelves in a random manner without some scheme of classification the library would be essentially useless. Similarly, if the plants of the earth were not classified it would be impossible to deal with them in any intelligent manner from a scientific standpoint.

Classifications are, of course, man-made and have no physical existence in nature as do the structures and processes of plants. There is, therefore, no single correct system of plant classification. Plants could be classified on the basis of a variety of criteria such as size, color, economic use, habitat, medicinal properties, or structural features, and as a matter of fact a number of such classifications have been made and used. However, ever since it has been recognized that species have arisen by evolution, and so are truly related to one another to a greater or lesser degree, biologists have attempted to make their basic classifications of plants and animals correspond with actual degrees of relationship as nearly as these can be determined. The plants in any of the major **divisions** of the **plant kingdom** are thus more closely related to one another than they are to plants in any of the other divisions. Each division is in turn divided into progressively smaller categories, at each level the most closely related plants being placed together in one group.

Structural similarities are the principal criteria used in determining the degree of relationship, but other criteria such as the number and kinds of chromosomes, the nature of the life cycles, and similarities in pigments and proteins present are also used. Systems of classification based on evolutionary relationships are **natural** systems, in contrast with **artificial** systems based on superficial similarities, *e.g.,* classifying pines and magnolias together because they are both evergreens or grouping roses and cacti because they are both spiny. Using a library analogy, the Dewey Decimal System is a natural classification since it groups books on the same subject, while classifications based on the size, weight, or color of cover of the books would be artificial.

The older systems of plant classification were essentially artificial, since the concept of evolutionary relationship had not yet developed. Even the famous system of the great Swedish botanist, Carolus Linnaeus (1707–1778), was artificial. During the past hundred years several natural systems of plant classification have been used by botanists, each successive one being based on increased knowledge of plant relationships and on the re-evaluation and re-interpretation of older information. There have been various modifications of each of the major systems, reflecting differences in personal opinion of various botanists. The systems of classification currently in use are by no means perfect, for our knowledge of plants is still far from complete and there is always room for differences in opinion and interpretation.

The Broader Classification of the Plant Kingdom. A number of different natural systems of classifying plants into major groups have been proposed and used. The system of Eichler (1883) divided the plant kingdom into four divisions: **Thallophyta** (algae and fungi), **Bryophyta** (mosses and liverworts), **Pteridophyta** (ferns, club mosses, and horsetails), and **Spermatophyta** (seed plants). This system was used by botanists for many years and is still used in some textbooks, but during the first part of the present century many botanists became dissatisfied with it, principally because it failed to recognize the great and basic diversity of the various groups of algae and fungi and because it made what seemed to be too major a distinction between the seed plants and other vascular plants (ferns, horsetails, and club mosses).

A system widely used at present (Tippo, 1942) divides the plant kingdom into twelve **phyla,** the first eleven being non-vascular and the last including all the vascular plants. The eleven phyla of non-vascular plants are the same as the eleven groups discussed earlier in this chapter. Some botanists feel, however, that this system is deficient in lumping all vascular plants into one phylum and all true fungi into another. The system of Bold (1956) divides the vascular plants into nine major groups, the fungi into three (phycomycetes, ascomycetes, basidiomycetes), and the bryophytes into two (liverworts and mosses). Altogether this system divides the plant kingdom into twenty-four major groups **(divisions),** the other ten being the same as those in the Tippo system. Many botanists feel that this is too many major groups, none of the systems being generally favored by all botanists. All the systems recognize the groups we have discussed as distinct ones, the different classifications merely representing diverse opinions as to the degrees of relationship among groups.

Smaller Categories of Plant Classification. Each of the major groups of plants includes many species of plants more or less distantly related to one another. To represent properly the varying degrees of relationship between plants, botanists have devised a classification hierarchy, each successive level consisting of smaller groups of organisms more closely related to one another. **Classes** are divided into related **orders,** the orders into related **families,** the families into related **genera** (singular, **genus),** and the genera into related **species.** Species may be regarded as the basic unit in the general scheme of classification, but in certain species with a considerable range of hereditary variation among its individuals **subspecies, races,** or **varieties** are distinguished, the latter being primarily a horticultural distinction. It is sometimes necessary to resort to subgroups at other than the species level to provide an adequate relationship grouping, so sometimes we find such categories as subclasses, suborders, and subfamilies.

The orders and smaller categories of classification are generally the same in one broad system of classifying the plant kingdom as another, being shifted as units when new divisions, phyla, or classes are established. However, no unit of classification can be regarded as fixed and unchangeable, though orders and families have been subject to less change with increasing knowledge than either the larger or smaller categories. Plant **taxonomists** (those botanists specializing in the classification of plants) are continually re-evaluating established genera and species and frequently decide on the basis of available evidence that what has been considered to be a single genus or species should be divided into two or more, or that several genera or several species should be combined into one. For this reason, and also because new species are continuously being discovered, it is impossible to give a precise number of plant species on earth.

One of the most difficult tasks in plant classification is the determination of the limits of a species, since species are generally not sharply distinct from other species in the genus. Such intergradations are to be expected in view of the evolutionary origin of species. Whether a species should be separated into two or more species or simply into subspecies or varieties, or whether two closely related species should be merged into one may involve very difficult decisions subject to differences of opinion among authorities on the group. Biologists even have difficulty in establishing a universally acceptable set of criteria or definition of a species.

A species is generally regarded as being a closely related group of organisms with numerous common structural features not present in related species, with a characteristic number of chromosomes, and with the capacity for interbreeding freely in nature. There may be exceptions, however, to these and other criteria used in limiting species. That a species may include a considerable variety of individuals is evident from the fact that all humans belong to a single species and all dogs to another.

Although classification is subject to continued change and improvement as well as to valid differences in opinion among authorities, the situation is not as fluid and chaotic as the preceding discussion may have implied. As international botanical code specifies certain procedures that must be followed in establishing new groups and in naming new species, thus providing a reasonable degree of stability that makes the system of classification useful in dealing with plants and referring to them by name, as well as sufficient flexibility to classification in line with our increasing knowledge of the relationships between various plants.

The common concepts of the scope of a category of classification are often quite different from the actual scope as defined scientifically. This is particularly true of families. For example, when a person who is not a botanist refers to the rose family he is likely to mean simply the different varieties or species of roses—in other words, the rose genus. The rose family actually includes strawberries, apples, pears, blackberries, spiraea, and many other genera as well as roses. The lily family includes onions, garlic, tulips, hyacinths, and other genera as well as lilies. Other examples of common genera from a number of different families are the oaks, maples, elms, amaryllis, goldenrods, and hollies. Each of these might commonly but incorrectly be referred to as a family.

Scientific and Common Names. Each group of plants at every level of classification has a **scientific name,** and frequently (but not always) one or more **common names.** Orders of plants rarely have common names. In our survey of the plant kingdom we used the common names of the major groups, though in a few cases they had to be synthesized from the scientific name since no truly common name really exists. In general, we shall use the common names of species in this book if they are widely used and precise enough.

The scientific name of a species consists of two parts. The first is the name of the genus and the second the name of the species, both always printed in italics with the genus name capitalized. In addition, the scientific name should include an abbreviation of the name of the botanist who named the species. Thus, the sugar maple is *Acer saccharum* Marsh. (for Humphrey Marshall), the scarlet maple *Acer rubrum* L. (for Carolus Linnaeus), and the Sierra maple *Acer glabrum* Torr. (for John Torrey). The genus and species names are always in Latin, though they may be only latinized forms derived from the Greek or some modern language. There is only one valid scientific name for each species, though it may have invalid scientific synonyms.

People sometimes wonder why it is necessary to use scientific names at all, but there are many good reasons. Common names can be given by anyone and are under no control, so one plant may have many common names or the same common name may be used for several different species. For example, there are said to be eighteen different species in the United States called "snakeroot," and on the other hand a single species may have a number of common names, such as the willow oak (*Quercus phellos* L.) that is also called pin oak, peach oak, water oak, swamp willow oak, red oak, and laurel oak, several of these names being used more widely for other oaks. Each language has its own common names for plants, and these are difficult to translate precisely, while the same scientific names are used throughout the world. When a species is referred to in a scientific publication it must be precisely identified, and usually this can be accomplished only by use of the scientific name.

Scientific names are a necessity rather than an affectation, and even in technical works botanists frequently use a common name of a species once they have identified it clearly by use of its scientific name. Finally, many of the more obscure species have no common name at all, or at the most just a common genus, family, or class name that is much too broad for scientific purposes.

Chapter 3

THE VEGETATIVE ORGANS OF FLOWERING PLANTS

Although the life processes of plants are basically very similar to those of animals, and although plants as well as animals are built up of cells, the organs of plants are not as numerous, complex, or specific in function as those of animals. In contrast with the lungs, stomach, liver, kidneys, brain, spleen, skeleton, and other organs of a higher animal, the roots, stems, and leaves are the only basic organs of a higher (vascular) plant. All other plant structures such as thorns, tendrils, or bulbs are simply modifications of one or more of these three organs. As you will discover in Chapter 12, even flowers, fruits, and seeds are modified derivatives of leaves. These reproductive organs will be considered in Chapter 4, the present chapter being devoted to the vegetative organs—roots, stems, and leaves. We shall consider both their external structure (morphology) and the details of their internal structure (anatomy), but first we must gain some understanding of the variety of cells that make up the bodies of vascular plants.

PLANT CELLS

In the algae and fungi some species are composed of only a single cell, and even in those species composed of dozens, hundreds, thousands, or millions of cells the cells are generally not very specialized and are all of the same few types. A greater degree of cell specialization occurs in the mosses and liverworts, but none of these lower plants has the great variety of cells present in vascular plants.

Meristematic Cells. The **meristematic** (or embryonic) cells are found in the tips of roots and stems, in the cambium between the wood and bark of trees and shrubs, and throughout an embryo plant.

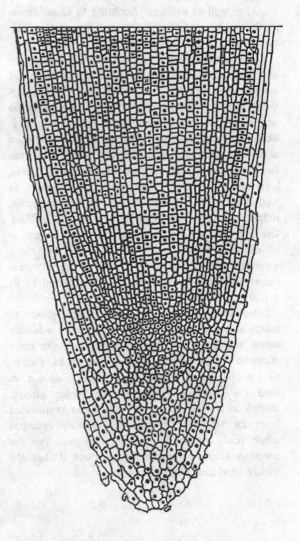

Fig. 29. A section through an onion-root tip as seen under a microscope. The larger cells at the bottom and along the sides are in the root cap, and above them are the small, cubical meristematic cells.

These cells provide the best material for beginning a study of plant cells since they are quite unspecialized and quite similar to what is sometimes called a "typical" cell whether of a plant or animal. The meristematic cells undergo repeated cell divisions, resulting in the increase in number of cells that is an essential part of growth, and then finally enlarge and become modified into all the many types of cells making up a vascular plant.

Meristematic cells can most readily be obtained for study from the root tips of a plant such as the onion that has been growing in water. The root tips are cut off, killed, fixed, cut lengthwise into very thin sections (after embedding in paraffin to make slicing easier), and mounted on a microscope slide and stained with dyes to make the various cell structures more visible under the microscope. Such a section of an onion root tip as viewed under the low power (X100) of a microscope is shown in Fig. 29. The walls enclosing each of the approximately cubical cells and the prominent spherical nucleus of each non-dividing cell are readily visible at this magnification. When a cell is viewed at high power (X430) many of the cell structures shown in Fig. 30 are visible.

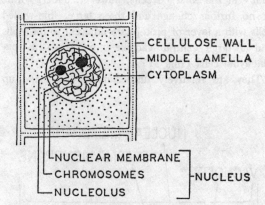

Fig. 30. A single meristematic cell such as those shown in Fig. 29, highly magnified.

The **cell wall,** though it may seem to be a single structure at low magnification, consists of two distinct layers: the **cellulose wall** completely surrounding a cell, and the thinner **middle lamella** (composed of pectic compounds) that occurs between the cellulose walls of adjacent cells and cements them together. It is the pectic substances in the middle lamellae of cells in fruits that cause jellies made from them to jell. As fruits become ripe the pectic compounds undergo changes that prevent jelling, and so ripe fruits can be used for jelly only if commercial pectins extracted from plant tissues (such as grapefruit skins or carrots) are added. The same changes in pectins of the middle lamella of ripe fruits result in the cells being held together less firmly, and so in the fruits becoming softer. The cellulose walls of meristematic cells, as well as the middle lamella, contain pectic compounds, interspersed among the rather loosely arranged cellulose fibers. Practically all plant cells are enclosed in a cell wall, but animal cells do not have cell walls. This constitutes one of the most universal differences between the two kingdoms of organisms.

Within the cell wall is found the **protoplast,** the tiny bit of living matter of the cell. The protoplast consists of two main parts: the spherical **nucleus** and the **cytoplasm,** the rather thick and viscous liquid (similar in consistency to raw egg white, but much more complicated in structure) that always completely surrounds the nucleus and is adjacent to the cell wall. The surface of the cytoplasm at the wall is somewhat denser and more coherent than the body of the cytoplasm and is known as the **plasma membrane.** This is comparable to the membrane surrounding an animal cell (the structure sometimes incorrectly called a "cell wall"). The cytoplasm includes several kinds of tiny organized bodies that are denser than the rest of the cytoplasm and somewhat gelatinous, each one surrounded by a membrane. Of these bodies only the **plastids** are large enough to be seen readily with an ordinary classroom microscope. Some plastids **(chloroplasts)** contain chlorophyll and carotenoids and are the site of the process of photosynthesis; others **(chromoplasts)** contain red, orange, or yellow carotenoid pigments but no chlorophyll. Plastids are generally somewhat oval or may be partially flattened ovoids.

Considerably smaller than plastids are the **mitochondria,** rod-shaped bodies that are seen to have a highly convoluted surface when viewed in electron microscope pictures. The mitochondria are the site of at least a good many of the important steps in the process of respiration. Still smaller than the mitochondria are the spherical **microsomes** (or **ribosomes)** that are arranged along the lamellar and rather irregular, convoluted, and branching **endoplasmic reticulum** of the cytoplasm. They are usually considered as a part of it. The microsomes are the site of protein synthesis; some cells also contain other specialized cytoplasmic structures. An electron microscope is essential for study of the structure of both the microsomes and the endoplasmic reticulum.

The knowledge of the minute structure of cells gained by the use of the electron microscope during the past decade or two and our increasing knowledge of cellular biochemistry have made at least two points increasingly clear:

1. The protoplast of a cell, including what used to be considered the rather unorganized cytoplasm, is highly organized even at submicroscopic levels.

2. Many if not all biochemical life processes are localized in specific cell structures rather than occurring throughout the cell.

The high degree of structural organization characteristic of the cytoplasm is also found in the nucleus. The nucleus of a cell is surrounded by a **nuclear membrane** and inside the nucleus are one or more smaller spherical bodies known as **nucleoli.** The exact function of the nucleoli is not known, but they apparently play a role in the transfer of genetic information from the nucleus to the cytoplasm. The **genes** (the hereditary potentialities or basic genetic information codes) are located in the chromosomes of the nucleus. In a non-dividing cell the chromosomes are long and threadlike and are visible as a tangled mass that was formerly interpreted as being a network. As a cell begins dividing, the chromosomes become shorter, thicker, and more distinct, permitting the number of chromosomes in a cell to be counted. Each species of plant or animal has a characteristic number of chromosomes in its cells. The body cells of animals and vascular plants contain two sets of chromosomes: one originally derived from the egg and one from the sperm that fertilized it and thus gave rise to the new individual. One consequence of this is that each cell contains two sets of genetic information or genes and another consequence is that the number of chromosomes in a body cell is characteristically an even number, 8, 12, 14, 16, 24, and so on, the number depending on the species. The chromosomes and nucleoli are immersed in a more liquid part of the nucleus known as the **nuclear sap.**

In the cytoplasm of older meristematic cells, a number of clear spherical areas of varying size may sometimes be seen. These are droplets of water with various sugars, salts, and other substances dissolved in them and are referred to as **vacuoles.** The vacuoles are surrounded by a cytoplasmic membrane known as the vacuolar membrane, and the solution making up the vacuole is known as the **cell sap.** The cell sap is not living. Other non-living inclusions in cells include crystals of proteins or of insoluble salts such as calcium oxalate and grains of starch, but such inclusions are generally not abundant in meristematic cells.

Parenchyma Cells. The older meristematic cells farthest back from the tip of a stem or root generally stop dividing and begin to increase greatly in size, generally growing much more in length than in breadth. Probably the first step in this cell enlargement is the increase in size of the cell wall by active deposition of new cell wall materials. There is also an increase in the amount of protoplasm in the cell. As the cell enlarges, more water enters the cell and much of this collects in the vacuoles, causing them to enlarge and then merge with one another. The final result is that one large central vacuole occupies most of the volume of the cell. Some of the cells may undergo further changes in size, shape, or structure and so develop into the various specialized cell types we shall consider later. Many of the enlarged cells, however, undergo little or no further change and these large, thin-walled cells with a layer of protoplasm adjacent to the walls and a large central vacuole are known as **parenchyma** cells (Fig. 31).

They are relatively unspecialized and make up

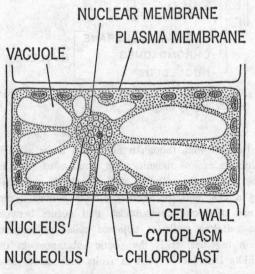

Fig. 31. Parenchyma cell.

a substantial part of the volume of a herbaceous plant and of the leaves and flowers and perhaps the fruits of woody plants. It is in the parenchyma cells of a plant that most of the important life processes occur. Parenchyma cells that contain chloroplasts, rather than other kinds of plastids, are sometimes called **chlorenchyma** cells. Some parenchyma cells are six-sided with about the same proportions as a shoe box, while others are fourteen-sided or some other shape. The shape of a parenchyma cell results to a large extent from the pressure of the adjacent cells, an isolated cell being essentially spherical.

Parenchyma-type Cells. A number of different cell types present in vascular plants have the general characteristics of parenchyma cells but differ from them in specific details. **Collenchyma** cells (Fig. 32) differ in having marked thickenings of their walls in the corners of the cells. **Epidermal**

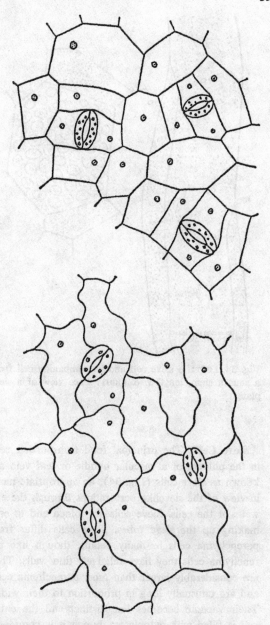

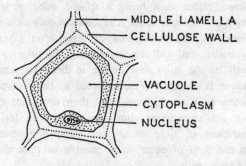

MIDDLE LAMELLA
CELLULOSE WALL

VACUOLE
CYTOPLASM
NUCLEUS

Fig. 32. Collenchyma cell.

cells (Fig. 33) make up the epidermis that covers herbaceous stems, roots, and leaves and are usually quite thin in comparison with their width. The epidermal cells of some species are quite regular in shape and may have the proportions of a rectangular or hexagonal ceramic tile, while those of other species look like pieces of a jigsaw puzzle. The epidermal cells of stems and leaves are also characterized by a layer of waxy substance **(cutin)** that covers their outer surfaces in contact with the air. The epidermal cells secrete this cutin, which is relatively waterproof. The layer of cutin is referred to as the **cuticle.**

Scattered among the ordinary epidermal cells of leaf and stem epidermis are pairs of **guard** cells (Fig. 33) bracketing a small pore through the epidermis. These guard cells are frequently bean-shaped, and they change shape as they gain or lose turgidity,

Fig. 33. Epidermal and guard cells of *Zebrina* (top) and sunflower (bottom).

closing up the pore when they are not turgid and opening it when turgid. A pore and its guard cells are called a **stomate** (or **stoma),** the stomatal pores providing channels through which oxygen, carbon dioxide, and water vapor molecules move between the leaf and the outside air. Another parenchyma-type cell, found only in the food transporting tissues (the **phloem)** of a plant is the **companion** cell, which will be considered in more detail later on.

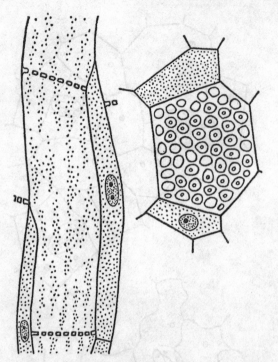

Fig. 34. LEFT: A sieve cell and its companion cell from a squash stem, lengthwise. RIGHT: Face view of a sieve plate.

Sieve Cells. The principal food transporting cells in the phloem of a vascular bundle or leaf vein are known as **sieve cells** (Fig. 34), an appropriate name in view of the sievelike perforations through the end walls of the cells. Sieve cells are joined end to end, making up the **sieve tubes.** Sieve cells differ from parenchyma cells in many details, though like parenchyma cells they have relatively thin walls. They are considerably larger than most parenchyma cells, and are unusually long in proportion to their width. Their vacuole becomes less distinct and the entire cell is filled with cytoplasm, though it is commonly less dense in the center than toward the sides. There is no nucleus in a mature sieve cell, the nucleus having disintegrated during development. Such nuclear disintegration is rare among cells and cells in which it occurs (the red blood cells of man and some other mammals) live only a few days or weeks in most cases. However, sieve tubes may live several years after losing their nuclei. The reason this is possible is not definitely known, but it is believed that the nuclei of the companion cells that are adjacent to the sieve cells may function for the sieve cells as well as their own cells.

Living and Dead Cells. All the cell types considered so far—meristematic, parenchyma, parenchyma-type, and sieve cells—are functional in a plant only when alive and carrying on their various life processes, with rare and minor exceptions. There are other kinds of cells in vascular plants, however, that function after they are dead and consist of nothing but cell walls. These cells, indeed, do not play their characteristic roles fully until after they are dead. They include cork cells, fiber cells, stone cells, and two kinds of water-transporting cells—the tracheids and the vessel elements.

Cork Cells. Cork cells found in the bark of woody stems and roots are characterized by having their walls impregnated with a waxy, waterproof substance known as **suberin.** The cells are generally box-shaped, commonly rather flat, and fit together tightly in most cases.

Stone Cells. Stone cells are usually about as wide as they are long and are somewhat irregular in shape, rather than being a sphere, cube, or other regular geometric form. They consist only of a very thick cell wall composed of layer after layer of secondary cell wall deposited inside the primary cell wall by the protoplast before it died. The central **lumen** is the space once occupied by the protoplast. Stone cells are less universal than the other types we are considering, but are responsible for the hardness of date seeds and the sandy granules in the flesh of pears, each granule being a cluster of stone cells.

Fibers. Fibers (Fig. 35) are long pointed cells with thick walls and a rather small lumen, and differ from stone cells principally in their shape. The two types of cells are commonly classified together as **sclerenchyma** cells. Fibers are quite abundant in the vascular tissue of angiosperm plants and function as mechanical supporting cells.

Tracheids. Like fibers, tracheids (Fig. 35) are long, pointed cells, but they have somewhat thinner walls and a larger lumen. Tracheids usually have perforations through their secondary walls, known as **pits,** and these are larger and more conspicuous than the pits of fibers. Some pits are simple, but those in the wood of pines, for example, are bordered pits that look something like doughnuts in face view. Tracheids play a dual role—they serve as tubes through which water flows and they also provide mechanical support. The wood of pines and other gymnosperms is composed mostly of tracheids.

Vessel Elements. In angiosperms water transloca-

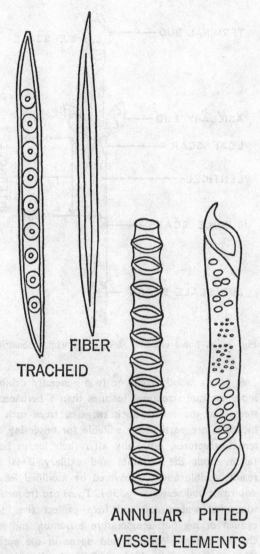

FIBER
TRACHEID

ANNULAR PITTED
VESSEL ELEMENTS

Fig. 35. All these types of cells function in a plant after they have died.

tion occurs principally through long tubes known as vessels (Fig. 35). A vessel may be as much as three or four feet long, though in some species they are only a few inches long. Each vessel is composed of a series of cells joined end to end with their end walls dissolved out. Each of the cells making up a vessel is called a **vessel element.** Vessel elements are commonly much larger in diameter than tracheids and have a considerable variety of secondary wall thickenings, ranging from rings through spirals to netted (reticulated) thickenings and to walls interrupted only by pits. The wood of angiosperm or hardwood trees is composed mostly of vessels and fibers, each playing

one of the two roles played by tracheids. There is evidence that both vessel elements and fibers arose from tracheids in the course of evolution.

Cell Size. After examining the drawings of cells as seen through the microscope you may need to be reminded that cells are minute objects. Parenchyma cells generally range between 0.01 and 0.1 mm. in width and may be two to three times this long. Thus there may be as many as half a million cells per cubic mm. In pith or the flesh of fruits, however, there may be parenchyma cells as much as 1 mm. in diameter, and these are visible without magnification by a person with good eyesight. Fibers usually range from 1 to 8 mm. in length, though in certain tissues of some species they may be 200 mm. or even longer. However, their diameters are small enough that they are visible only under the microscope.

In general, the cells of vascular plants are relatively large—considerably larger than most cells of animals and so much larger than bacterial cells that a parenchyma cell could hold thousands of bacteria. Blue-green algae also have very small cells. Probably the largest plant cells are those of the marine green alga *Valonia* (Fig. 10) that are up to an inch or more in diameter. Most cells, however, are so small that they cannot be seen without a microscope.

STEMS

As has been mentioned earlier, stems, roots, and leaves are organs present only in vascular plants although some of the lower plants have structures that superficially resemble one or more of these organs. The stem seems to be the most basic of the three organs, since the primitive vascular plants known as psilophytes (Chapter 2) include species with stems but no leaves or roots. The stems of the various species of vascular plants have a wide range of structures and in some cases are hardly recognizable as stems. However, all stems have certain structural features in common (excluding, for some characteristics, some of the psilophytes). These include the presence of three fundamental tissue systems that make up the stem (the **tegumentary** or covering tissues, the **vascular** tissues, and the **fundamental** tissues internal to the first and surrounding the second), the presence of **terminal buds** and regularly arranged **lateral buds,** and the bearing of leaves and their modifications. The lateral buds are normally borne on the stem in the axil of a leaf (the angle formed by the leaf base and the stem) and so are referred to as **axillary buds.** The point of attachment of a leaf and its axillary bud is called a **node,** and the region of a stem from one axillary bud to the next is an **internode** (Fig. 36). Nodes and internodes are strictly stem characteristics, never being present on either leaves or roots.

One obvious way of classifying stems is as **woody** and **herbaceous.** The latter lack the hard texture and firm mechanical support provided by the cylinder of wood and the corky bark of woody stems and are rather soft, commonly green, and covered by an epidermis a single cell thick. Perennial plants that live year after year generally have woody stems while annual or biennial plants generally have herbaceous stems, but there are exceptions in both cases. Despite the rather obvious differences between the two types of stems, the differences are really not fundamental or of any basic significance in the natural classification of plants. A single family may include trees such as apples and cherries, shrubs such as roses and spiraea, and herbs like the strawberries. There are woody as well as herbaceous species of tobacco. Furthermore, the stems of some herbaceous plants such as the tomato, sunflower, and tobacco become semiwoody by the end of the growing season. A young woody stem is basically the same structure as a herbaceous stem. We shall, however, find it useful in our study of external and internal stem structure to distinguish between woody and herbaceous stems.

External Stem Structure. Perhaps the most obvious structural features of most herbaceous stems are the nodes, with their axillary buds, and the attached leaves. The green color of most herbaceous stems comes from chlorenchyma cells located under the transparent epidermis covering the stem, not from the epidermis itself. The stems of some species are quite smooth while those of others are hairy or sometimes prickly. Though most stems are round in sectional view, the stems of members of the mint family are square while those of sedges are triangular. Some species of cactus have flattened and leaflike stems, the true leaves being spines. The terminal buds of herbaceous stems are generally rather small and not too easy to see because they are surrounded by the young leaves just beginning to enlarge.

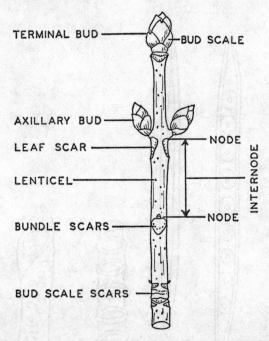

Fig. 36. A horse chestnut stem in its winter condition.

A young woody stem or twig generally exhibits more external structural features than a herbaceous stem, and the rather thick twigs of trees such as hickories are particularly suitable for observing external structures, especially after their leaves have fallen. Both the terminal and axillary buds are readily visible and are covered by modified leaves known as **bud scales** (Fig. 36). These are frequently somewhat waterproof and help protect the buds against drying out, temperature extremes, and mechanical injury. When a bud opens in the spring the bud scales fall off, but others develop by the completion of the year's growth. Buds are present on trees and shrubs throughout the year, contrary to the common misconception that they arise anew in the spring.

If a longitudinal section is made through a bud its nature becomes obvious (Fig. 37). A bud is simply the young stem tip covered by the small young leaves and sometimes by bud scales, too. At the very tip of the stem (the **apex**) are the meristematic cells that by repeated cell divisions provide the cells that increase the length of the stem and also the cells that develop into the axillary buds and leaves. The young stem within a bud has very short internodes. As a bud grows into a mature stem the internodes elongate greatly, separating the leaves from one another, and the leaves and axillary

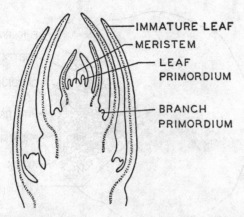

IMMATURE LEAF

MERISTEM

LEAF
PRIMORDIUM

BRANCH
PRIMORDIUM

Fig. 37. Lengthwise section through a bud of honeysuckle (enlarged).

buds rapidly grow to their final sizes. All the leaves that will appear on a tree in the spring are present within the buds during the previous winter. The axillary buds develop into the branches, but in woody plants the buds of the current year's growth rarely grow into branches and the buds may not begin growing into branches until they are two or more years old. Some axillary buds never grow into branches even though they remain alive and are capable of doing so. If all axillary buds developed into branches the first year they were formed, a tree or shrub would be an extremely dense and almost impenetrable thicket of branches.

Some buds contain young flowers instead of leaves, or flowers along with some leaves. The difference between a flower bud and a leaf bud is not as marked as one might suppose. A flower really consists of a stem bearing modified leaves (the sepals, petals, stamens, and pistils). However, the extremely short internodes between these modified leaves never elongate, there are no axillary buds, and there is no further terminal growth of the stem of the flower.

Just underneath the axillary buds of a woody stem that has lost its leaves are the **leaf scars** (Fig. 36), representing the point of attachment of the leaves to the stem. Within the leaf scar are small dots, the **bundle scars,** where the vascular bundles entered the leaf stalk. The shape of the leaf scar and the number and arrangement of bundle scars are characteristic of a species and are useful in identifying woody plants in their winter condition. The leaves and axillary buds of some species are located on opposite sides of the stem, the pair at

the next node being either directly below these (two ranked) or 90° around the stem (four ranked). In contrast with such **opposite leaves** are the **alternate leaves** of other species, which have only one leaf (and bud) at a node. A string extending from one leaf base to the next one would describe a spiral around the stem, and the pitch of this spiral is characteristic of a species.

At the base of the last year's growth of a stem there is a circle of **bud scale scars,** making it possible to determine just how much the twig grew in length the previous season. Sometimes the bud scale scars of the previous two or three years are also still visible. It should be noted here that all the growth in length of stems occurs in the part of a stem that emerged from a bud in the spring of the year. Once the growing season is over, that part of the stem will never get any longer, though it will continue to increase in diameter year after year in a woody plant. Thus, the common misconception that a tree trunk keeps growing in height at its base and that a nail driven in a tree trunk will be raised higher and higher above the ground as a tree grows has no basis in fact.

Very young woody stems may be green and have an epidermis just like herbaceous stems. However, even during the first year, the twig grows in diameter. As the outer tissues split, corky bark develops, first in patches and then over the entire stem surface. Visible on the bark of young twigs of most species are dots or streaks. These are the **lenticels** (Fig. 36), specialized regions where cells are loosely arranged and through which gases may pass between the stem and the outside air. The lenticels are not readily visible on the older, thicker stems of many species, but the elongated lenticels on the trunks of birch and cherry trees and a few other kinds are very obvious and striking features.

As a woody stem grows in diameter year after year the surface features such as leaf scars and bud scale scars at first become stretched and then obliterated as the bark keeps cracking and splitting because it is no longer large enough in circumference to go around the enlarging stem. While the bark gets somewhat thicker as a stem grows older, it never gets as thick as the wood. One reason is that each year a thicker layer of wood than bark is produced and another reason is that the older, outer layers of bark keep falling off. The bark of many species of trees becomes deeply ridged as it splits, but in a few species such as birch, cherry, and sycamore the bark splits off in sheets rather than in furrows. The bark of any species of tree has a characteristic color as well as characteristic structure, and these bark characteristics can be used along with twig characteristics in identifying a tree in the winter.

Another thing characteristic of a species is its pattern of branching, which determines the general form of the plant. Some trees like pines and firs have terminal buds that keep on growing throughout the life of the tree, resulting in a main trunk visible all the way to the top, with all the main branches extending out from it. Other trees such as oaks, elms, and maples have terminal buds that continue growth for a limited period of years and then die. Axillary buds then take over, resulting in branching and the loss of a single main trunk. This occurs repeatedly, resulting in a much-branched crown.

Shrubs differ from trees, not only in their generally smaller size, but also because they begin branching at or near ground level and so have no main trunk at all. The characteristic appearance of each of the various herbaceous species is also greatly influenced by their pattern of branching. Some species like sunflowers and corn have few or no branches, while others such as beans and marigolds are highly branched.

Internal Stem Structure. If a thin cross section of a herbaceous dicot such as the bean plant is prepared for observation under the low power of a microscope, it is found to be composed of a limited number of rather distinct and characteristically arranged tissues (Fig. 38). The outermost one is the **epidermis,** which is only one cell thick in most species. The waxy cuticle covering the epidermis and secreted by the epidermal cells is usually visible in stained sections. Inside the epidermis is the **cortex,** usually about five or six cells thick. The cortex of most herbaceous stems is composed principally of chlorenchyma cells that carry on photosynthesis, but there may also be some collenchyma cells in some species and perhaps some plain parenchyma cells. In the center of the stem are found the parenchyma (or more rarely chlorenchyma) cells that make up the **pith.** These cells are generally unusually large. In some species the central part of the pith breaks down, resulting in a stem with a hollow center. Between the pith and the cortex

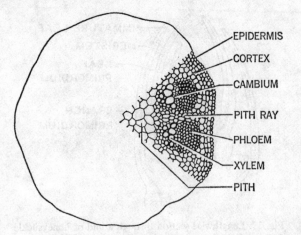

Fig. 38. Cross section of a dicot stem.

is a circle of **vascular bundles,** separated from one another by extensions of the pith and cortex (the **rays).**

The outer part of each vascular bundle is made up of the **phloem,** or food-conducting tissue, while the inner and generally larger part is composed of **xylem,** or water-conducting tissue. The bundles of a few species have phloem on the inner edge of the xylem as well as in the usual position. Some species also have a distinct cap of fiber cells between the phloem and the cortex, this tissue sometimes being called the **pericycle.** The vascular bundles are sometimes referred to as fibrovascular bundles, particularly when they include pericycle fibers. Unlike the other tissues of herbaceous stems, both the xylem and phloem are composed of a considerable variety of different cells.

The principal cells of the xylem are the large vessels through which water flows up the stem, but the xylem also commonly contains some parenchyma cells and fibers. The most important cells in the phloem are the sieve cells through which food flows, but adjacent to each sieve cell are one or two companion cells. In addition, the phloem may also contain parenchyma cells and fiber cells.

In some species of herbaceous dicots there is a single layer of meristematic cells between the xylem and phloem. This is the **cambium,** and as a result of its cell divisions in a radial direction additional xylem is produced on its inner side and more phloem toward its outer side. This is secondary xylem and phloem, in contrast with the primary xylem and phloem that had its origin in the apical meristem of the bud. In a few herbaceous species a cambium (the **interfascicular cambium**) may also appear, bridging the rays and connecting the cambium of the adjacent bundles. The result is a continuous cylinder of cambium around the stem. If an interfascicular cambium is present, it may produce secondary xylem and phloem that merges with that of the bundles. However, even in the herbaceous species that do have a cambium, the amount of secondary xylem and phloem produced is not too great and the stem becomes only semiwoody as a result of the cambial activity.

Though the vascular bundles of a stem may be quite distinct from one another in a cross-sectional view, they are actually connected with one another sidewise, branching and merging like the wires in a netted wire fence or like the strings in a net. Thus, when the vascular system is visualized from a three-dimensional standpoint it is seen to be a hollow cylinder interrupted by the rays, with the pith inside this cylinder and the cortex outside it.

The stems of herbaceous monocots generally contain the same types of cells and tissues as those of herbaceous dicots, but they are arranged somewhat differently. In corn stems there are several vascular cylinders, one inside another, so the vascular bundles appear to be scattered throughout the stem and no definite pith is evident, even though the parenchyma tissue surrounding the bundles is commonly referred to as the pith. It is also difficult to identify cortical parenchyma as distinct from that of the pith, and so the term **ground parenchyma** is used to cover both. However, the outer regions of the cortex are distinct in containing both fiber cells that are responsible for the rigidity of the stem and chlorenchyma cells that give it a green color. As in dicots, the phloem is toward the outer side of the bundles, but the arrangement of the cells in the xylem and phloem of corn bundles is somewhat different and quite distinctive. The stems of some other herbaceous monocots are more similar in appearance to those of dicots, and a good many species have a hollow pith. Cambium is not present in monocot bundles.

A cross section of a young woody stem before any secondary growth has occurred reveals the same tissues present in a herbaceous dicot stem (Fig. 39). The principal differences are in their proportions, rather than in their positions. The pith and cortex are commonly relatively smaller in diameter than in a herbaceous stem, the rays are narrower, and

the xylem is generally thicker, though there is considerable variation from species to species in the proportions. A cambium is present in all woody stems, in contrast with its absence in some herbaceous species. Finally, pericycle fibers are more universal in woody than in herbaceous species. However, there are no really fundamental differences between a herbaceous dicot stem and the primary tissues of a woody dicot stem. The distinctive and marked structural differences between a herbaceous stem such as that of a bean plant and an old woody stem such as the trunk of an oak tree are primarily the result of the secondary growth of the latter.

Secondary Tissues of Woody Stems. Even during the first year of its life a woody stem acquires some secondary xylem and phloem as a result of the activity of its cambium. By the time a stem is a month or so old a cylinder of cells in the outer part of the cortex (the **cork cambium**) becomes meristematic, and then begins to produce cork cells toward the outside. When the layers of waterproof cork cells become complete they isolate the outer layers of cortical cells and the epidermal cells from a supply of water and food, and these cells then die. The result is that the stem is now covered with a layer of cork rather than its original epidermis (Fig. 39). This cork is also a secondary tissue.

Meristematic activity of both the vascular and cork cambiums generally ceases during the winter, but is resumed the following spring. During the second year new layers of secondary xylem and phloem are added by the cambium, which also increases the length of the rays. The resulting increase in stem diameter causes the cork and cortex to split at various points and then new cork cambium appears deeper in the cortex and produces more cork that covers the split area. As additional layers of xylem and phloem are laid down in successive years the new and deeper corky layers finally isolate all of the cortex and begin developing in the older and outer layers of the phloem. Thus, the cortex as well as the epidermis is lost and a true **bark** consisting only of phloem and cork tissues has developed. Each year new rings of xylem and phloem are added by the cambium, which moves outward as it lays xylem down behind it. Since its circumference is thus increasing year after year it is evident that the cambium itself is growing. This is accomplished by occasional cell division in

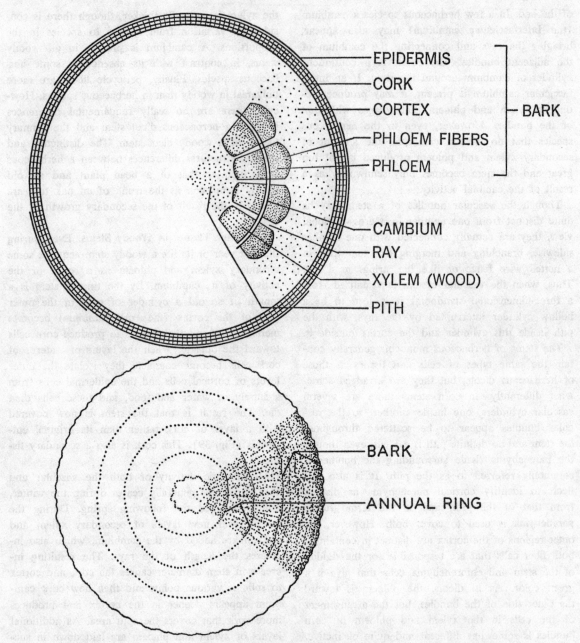

EPIDERMIS
CORK
CORTEX   ⎫ BARK
PHLOEM FIBERS
PHLOEM

CAMBIUM
RAY
XYLEM (WOOD)
PITH

BARK

ANNUAL RING

Fig. 39. TOP: Diagrammatic cross section of a one-year-old woody stem. BOTTOM: Cross section of a three-year-old linden or basswood tree. Note the annual rings in the xylem (wood) and the bark, composed only of phloem and cork. The top drawing is more highly magnified than the bottom one.

a tangential direction, in addition to the many cell divisions in a radical direction that produce the new xylem and phloem. Since the phloem and cork are, however, not able to grow tangentially they keep on splitting and new cork cambia appear deeper in the phloem. Thus the older layers of the phloem are isolated by cork, die, and finally fall off along with the older layers of cork. While a stem contains all the layers of xylem it has produced, only the phloem produced in the more recent years remains as part of the bark and of these only the younger layers of phloem are still active in transporting food.

A mature woody stem (Fig. 40), then, consists of the following tissues from the center outward: 1. the pith, which is strictly a primary tissue and never gets any larger than it was the first year; 2. the successive annual rings of xylem (the wood), from the oldest one next to the pith to the youngest one next to the cambium; 3. the cambium; 4. the phloem, with the youngest layer next to the cambium, and mixed with the older layers of phloem; 5. the cork cambium and cork.

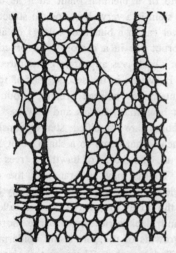

Fig. 40. Aspen wood portion, highly magnified.

The foregoing description of the structure and secondary growth of woody dicot stems applies equally well to the woody stems of gymnosperms such as pines and spruces. There are, however, important differences between dicots and gymnosperms as regards the composition of the xylem or wood. In dicots the water conducting tubes are vessels composed of vessel elements joined end to end, and fibers are present in the xylem along with the vessels. In gymnosperms the tracheids serve as both water conducting tubes and as mechanical support, thus performing the functions of both vessels and fibers. The wood of a pine tree is composed almost entirely of tracheids, the only other cells being the parenchyma cells of the pith rays, the pith, and the resin ducts. The latter are long tubes lined with parenchyma cells that produce and secrete into the ducts the rosin that is characteristic

of pines. (The term "rosin" is generally used to refer to the resin of the pine tree.)

In both the woody dicots and gymnosperms the **annual rings** of xylem are quite distinct from one another. The reason is that the vessels or tracheids produced during the rapid growth of spring are large and thin-walled, while those produced during the summer are smaller and have thicker walls. There is commonly a gradation between the spring and summer wood of a season, but growth of the annual ring ends abruptly at the close of the growing season. The large, thin-walled cells produced the following spring contrast sharply with the small, thick-walled cells formed the previous summer, thus setting apart the rings. Annual rings are commonly used in determining the age of trees and have also been used in determining past climatic conditions since wider rings are produced in wet years. Tree rings reflect the rather regular cycles of wet and dry years, and so have been used in dating prehistoric Indian pueblos in the Southwest by comparing tree-ring patterns from recently cut old trees with the patterns in the wooden beams of the pueblos. Since surveyors have often blazed trees to mark boundary lines, sections of tree trunks have frequently been used as court evidence as to which of several surveys was the first and as to the exact location of the survey lines. The blazes eventually heal over and become covered by new wood and bark, but a section through the tree shows what the last ring present at the time of blazing was and counting of the rings makes possible determination of just when the tree was blazed.

There are, however, certain precautions that must be observed in reading tree rings. Occasionally two rings are formed in a single year, as when a late freeze kills all the leaves on a tree soon after they have emerged. Then cambial growth stops for a while and resumes later when a new crop of leaves emerges from the buds. The result is two unusually narrow rings, that can usually be spotted. Similar pairs of narrow rings may be formed in a year if all the leaves have been stripped from a tree by an unusual infestation of caterpillars or other insects. Tree rings much wider on one side of the trunk than the other suggest a local environmental difference, such as shading of one side of the tree by larger trees, thus reducing the photosynthetic production of food on that side and consequently the growth rate. If the rings of a tree suddenly become much wider than they were it may mean that the tree had been growing in the shade of other trees in a forest, and that the neighboring trees were cut down for lumber, thus giving the tree in question more light.

Modified Stems. We usually think of stems as more-or-less upright aerial plant organs that are elongated, self-supporting, and bear leaves and flowers. Most stems meet this description but there are various kinds of modified stems (Fig. 41) that fail to meet one or more of these qualifications.

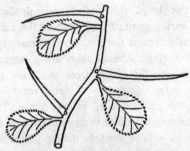

THORNS (HAWTHORN)

TUBER (POTATO)

RHIZOME (SOLOMON'S SEAL)

Fig. 41. Some types of stem modifications.

Some stems are prostrate on the ground and grow horizontally. Among these are trailing vines and the **runners** or **stolons** of various grasses, strawberries, and some other plants. **Rhizomes** are horizontal stems that grow underground and are frequently rather fleshy. All ferns that grow in temperate climates have rhizomes rather than erect, above-ground stems, though some tropical ferns (the tree ferns) have erect stems. Iris is another plant with rhizomes, only the leaves and flower stalks emerging from the ground. **Tubers** are much enlarged portions

of rhizomes, so the tubers of Irish potatoes are really stems, even though they grow underground. (Sweet potatoes are really roots.) The "eyes" of a potato are really axillary buds arranged at definite nodes, the leaves being reduced to tiny flaps of brown tissue that look like the potato skin.

Corms, such as those of the gladiolus, are also underground stems, but they have a vertical axis even though they are wider than they are high. The **bulbs** of onions, tulips, and other plants are also underground structures and are essentially large buds (as are heads of cabbage and lettuce). The actual stem tissue is the solid cone at the base, the rest of the bulb being overlapping leaves. During the first year of life of a biennial plant such as carrot or celery the stem is a very short conical structure near ground level (as in a bunch of celery) or at the top of the taproot (as in a carrot) with practically no internodes. The leaves appear to come directly from the roots. However, in the second year of the plant's life the internodes enlarge greatly, producing tall stems that bear many leaves and flowers.

True **thorns** are modified stems, consisting of short branches that come to a sharp point lacking a terminal bud. The thorns of hawthorn trees are good examples. Thorns may be branched, the outstanding example being the large and extensively branched thorns of the honey locust tree. The so-called thorns of roses and blackberries are not true thorns, being local outgrowths of the cortex and epidermis rather than modified branches. They are technically **prickles.**

In some species the leaves are very small and lack chlorophyll, their photosynthetic function being assumed by the stems, as in horsetails and cacti. In a few other species there are more-or-less regular stems with much flattened branches that look greatly like leaves but are seen to be stems on close inspection.

Stems as Means of Vegetative Propagation. Although the basic means of reproduction of vascular plants is sexual, involving propagation by seeds in the gymnosperms and angiosperms, plants may also be propagated by detachment of vegetative organs, either naturally or by man. Of the three vegetative organs, stems are the most widely used for vegetative propagation. Vegetative propagation depends on the capacity for regenerating missing parts. Thus a twig cut from a tree has all the parts of a complete plant except roots. By placing the twig in water or moist sand, and perhaps treating it with a root-inducing plant growth hormone, adventitious roots may be induced to form. We now have a complete plant that may grow into a large tree just like the parent. Such stem **cuttings** (Fig. 42) are one common means of vegetative propagation, most of the cultivated shrubs being propagated in this way. **Layering** is a modified form of cutting propagation in which long, drooping stems (such as those of raspberries) are placed underground near their tips. After roots form, the end of the branch is cut off and the new plant is dug up and planted. This is also a natural means of propagation, resulting in the tangled and almost impenetrable briar patches of wild raspberries and blackberries.

Tubers, rhizomes, bulbs, and corms all serve as both natural and horticultural means of propagating the plants that produce them. Potato tubers are cut into pieces before planting, thus providing many more plants from a lot of tubers than if each one were left intact. This is essentially a type of stem cutting. The rhizome of an iris or other plant may be cut into pieces, each piece giving rise to a new plant. Bulbs and corms produce small branch bulbs or corms that can be detached and used for propagation.

Apples, cherries, pears, and other fruit trees (as well as roses and some other woody plants) are commonly propagated by **grafting.** This method is used when cuttings do not root well, or when the root systems that develop are not strong. Branches (the **scions**) are cut from the plant to be propagated and are then grafted onto the stems of young plants of a related species from which the top has been removed (the **stock**). Apple scions are usually grafted onto crab apple stocks and roses onto wild rose stocks. The root systems of these stocks are more vigorous and disease-resistant than the roots of the cultivated varieties. By use of interlocking cuts of several types, stocks and scions of the same diameter, and waxed string and wax to bind the graft union and protect it from damage or drying out, it is possible to produce a successful graft union.

It should be noted that, contrary to popular belief, grafting is not a means of crossing or hybridizing plants. Both the stock and scion continue to have all the characteristics of the plants from which they came, no hereditary changes being brought about. This explains the thing that has puzzled many people: A rosebush that has been producing beautiful hybrid tea roses one spring bears nothing but wild roses.

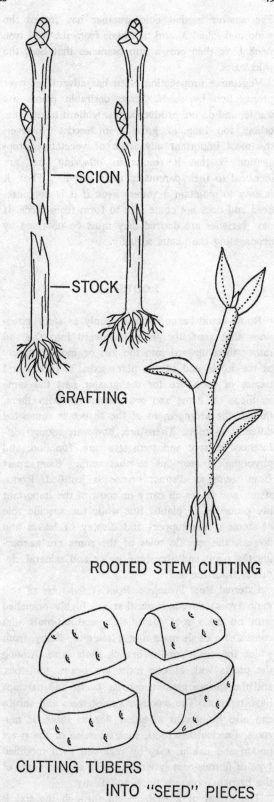

Fig. 42. Vegetative propagation by means of stems.

The answer is that cold weather has frozen the scion and killed it, and that buds from the wild rose stock have then grown into branches that bore the wild roses.

Vegetative propagation often has advantages over propagation by seeds. Some desirable plants are sterile, and do not produce seeds, while it takes some others too long to grow from seeds. However, the most important advantage of vegetative propagation is that it results in offspring that are identical to their parents in their heredity. Thus, it is easy to maintain a variety even if it is not purebred and does not come true to form from seeds. If new varieties are desired they must be obtained by propagating the plants sexually.

ROOTS

Roots could be considered simply as stem extensions that generally grow downward into the soil rather than upward into the air, or more properly as the lower end of the plant axis. The cells and tissues of roots are for the greater part the same as those of stems and are continuous with them, though the arrangement of the tissues is somewhat different in roots. There are, however, enough differences in root and stem structure, function, and physiological reactions so that setting them apart from stems as distinct organs is justified. Roots, stems, and leaves all carry on most of the important life processes of plants. But while the specific role of stems is the support and display of leaves and flowers, the specific roles of the roots are anchoring the plant and absorbing water and mineral elements from the soil.

External Root Structure. Root systems are of two main types: **fibrous roots** that are highly branched with no main root identifiable, and **taproots** that consist of a single main root, often quite fleshy, from which the much smaller branch roots arise. Among the plants with fibrous roots are beans, tomatoes, and the various grasses; carrots, beets, and parsnips have taproots. The woody roots of trees and shrubs can also be classed as either fibrous roots or taproots. **Fasciculated** roots, such as those of the sweet potato and dahlia, may be regarded as a modified type of fibrous-root system in which at least some of the branch roots are thick and fleshy.

The extent of a root system through the soil is often not appreciated, the root systems of most species occupying a very large volume of soil. The roots of most trees extend at least as far from the trunk as do the tips of the branches. The absorption of water and mineral salts occurs mostly near the ends of the roots, and so in watering or fertilizing a tree the applications should be made near the crowns, not next to the trunks. The absorbing surface of roots is greatly increased by the numerous root hairs found near the tips of most roots. These root hairs are long extensions of some of the epidermal cells and are in intimate contact with the soil particles. A single rye plant has been estimated to have over fourteen billion root hairs with a combined surface area of over four thousand square feet, and to this should be added the absorbing surface of the rest of the root epidermis. It is apparent that the root absorbing surface is far greater than the total leaf and stem surface exposed to the air.

Perhaps the most striking external structural difference between roots and stems is that roots do not bear leaves or axillary buds and therefore have no nodes or internodes. In contrast with the rather regular branching pattern of stems the branching of roots appears to be quite haphazard, though it is sometimes possible to note that the branch roots occur in a definite number of vertical rows (such as three or four). The reason for this will become apparent later. Since roots do not bear leaves (or modified leaves such as bud scales), the tip of a root is exposed, in contrast with the stem tip. However, most root tips have a **root cap,** composed of parenchyma cells. As the root grows through the soil the cells of the root cap are rubbed off, but more root-cap cells are added by the meristematic growing tip back of the cap. Woody roots, like woody stems, are covered with bark consisting of phloem and cork, but the tips have an epidermis like those of herbaceous roots.

Internal Root Structure. A section across a dicot root at the root-hair zone shows the similarities and differences between the internal structure of roots and stems. As in stems, the epidermis is generally only one cell thick, but there are no stomata and some of the epidermal cells bear root hairs (generally at a regular interval, as every third cell). Also, the root epidermis is not cutinized like the stem epidermis. The cortex of roots is composed of parenchyma cells and is generally thicker in proportion than that of stems, as well as being more precisely limited on its inner side by the presence of the single-cell-thick **endodermis,** a tissue not present in stems. The cells of the endodermis are of a general parenchyma type, but are modified by various degrees of suberization or thickening of the walls. The suberization probably plays some role in regulation of water movement across the endodermis.

Inside the endodermis is a pericycle, generally composed of a few layers of parenchyma cells, and inside this the vascular tissue. These tissues inside the endodermis are frequently referred to as the **stele.** One striking difference between the roots of many dicots and their stems is that there is no pith in the roots, the center of the root being occupied by xylem. There are usually outward extensions of the central xylem, the number of extensions—commonly three, four, or five—being characteristic of a species. Thus, a cross section of the xylem may have the appearance of a cross, star, or cogwheel. The separate bundles of phloem are located in the indentations of the xylem, quite a different arrangement than in the vascular bundles of the stem. At the junction of the root and stem the vascular tissues undergo complex interchanges that connect the root and stem patterns of organization.

Whereas the branches of stems arise superficially from buds, branch roots arise internally, originating in the stele. After the meristematic tip of the branch root has been differentiated from pericycle cells, it grows through the cortex and epidermis to the outside. The tissues of branch roots are arranged the same as those of the main root, and their vascular tissues join those of the main root. Branch roots of small diameter are sometimes called hair roots, and it should be noted that these are quite a different thing than root hairs.

Secondary Growth of Woody Roots. The secondary growth of woody roots follows about the same pattern as the secondary growth of woody stems and, except that no pith is present, the final tissues are the same: the xylem or wood, the cambium, and the bark composed of phloem and cork. However, there are two main differences in the initial stages of secondary growth. The first results from the different arrangement of primary xylem and phloem in the root. At first the cambium appears only between the bundles of phloem and the xylem, laying down secondary xylem behind it and secondary phloem in front of it. Thus, the indentations in the primary xylem are filled in with secondary xylem and a cylinder of xylem results. The cambium now becomes continuous around the circumference of

the xylem and from this point on the secondary growth is like that of stems. The annual rings of xylem in roots, however, are not as clearly defined as they are in stems. The second difference in secondary growth is that the cork cambium originates in the pericycle and as the layer of cork it produces becomes complete the tissues outside it (the endodermis, cortex, and epidermis) are all isolated and sloughed off. Eventually new cork cambiums form in the older phloem and so all the pericycle is lost, too.

Modified Roots. Modified roots are less numerous as to kinds and the modifications are less marked than some of those of stems. Although most roots accumulate some starch or other food, the fleshy taproots of plants such as carrots and beets and the fleshy fasciculated roots such as those of the sweet potato may be regarded as modified roots particularly adapted to food accumulation. Such roots are widely cultivated by man as food sources, sugar beets being the second most important commercial source of sugar.

Adventitious roots are those that grow from an organ other than the primary root system, principally from stems. In addition to adventitious roots that develop on cuttings, some species naturally bear adventitious roots. The prop roots of corn plants that arise on the stem above ground are adventitious (Fig. 43), as are practically all the underground absorbing roots of this species, the original primary-root system being only temporary. The roots by

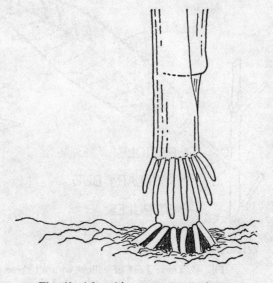

Fig. 43. Adventitious prop roots of corn.

which poison ivy clings to trees or other objects are adventitious, as are most other aerial roots of other species. The haustoria of the parasitic dodder plant that penetrate the host plant on which it is living and absorbs water, food, and minerals from it are specially modified aerial adventitious roots. After a dodder plant is attached to its host the true roots and lower stems die.

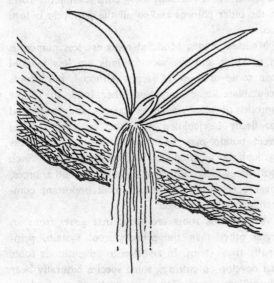

Fig. 44. Aerial roots of an epiphytic orchid.

The aerial roots of **epiphytes** (plants that are perched on another plant but do not obtain food or water from it), such as some of the tropical orchids, however, are true roots and not adventitious (Fig. 44). They absorb water from rain and dew and mineral salts from dust particles in the air. Some epiphytes such as Spanish moss rarely have roots, the leaves and stems absorbing the water and salts from the air. The aerial roots of many species contain chlorenchyma cells and so can carry on photosynthesis.

Roots as Means of Vegetative Propagation. Roots are a far less important means of vegetative propagation than stems, probably because roots produce adventitious buds less readily than stems do adventitious roots. However, sweet potato roots readily form adventitious buds and the stems that develop from them (the slips) are removed and planted, this being the standard way of propagating sweet potatoes. If these roots were left in the ground they would provide a natural means of vegetative propagation. The roots of dahlias do not form adventitious buds, and so can be used for propagation only if a piece of stem is attached. Several woody species, particularly shrubs, can be propagated by making cuttings of their roots. The roots of many hardwood trees develop adventitious buds after the

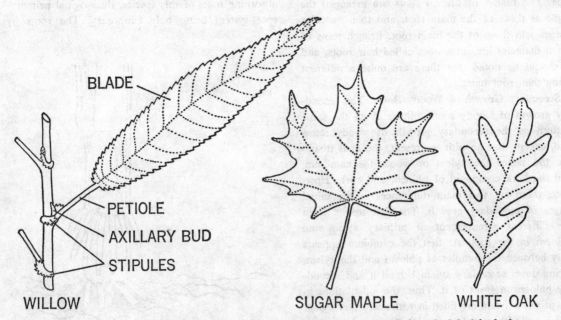

Fig. 45. LEFT: Leaf of willow with all three parts present. CENTER: Palmately lobed leaf of sugar maple. RIGHT: Pinnately lobed leaf of white oak. All three leaves are simple.

BLADE

PETIOLE

AXILLARY BUD

STIPULES

WILLOW   SUGAR MAPLE   WHITE OAK

tree has been cut down, and the sprouts that grow from these buds may eventually grow into large trees. However, unless the sprouts are severed from the parent this is a matter of regeneration rather than propagation.

LEAVES

Leaves are characteristically broad, thin structures well adapted to their primary role in the economy of a plant—the production of food by photosynthesis. There are, however, many modifications of this basic structural pattern, notably the needle-shaped leaves of pines and other gymnosperms. The many more extreme leaf modifications will be considered later.

External Leaf Structures. Leaves have three main parts: 1. the flattened leaf **blade** (or **lamella**); 2. the leaf stalk or **petiole**; 3. the paired **stipules** at the base of the petiole (Fig. 45). However, the leaves of many species lack one or two of these parts, while in other species they may be so highly modified as to be almost unrecognizable. If a leaf lacks a petiole and its blade is attached directly to the stem it is said to be **sessile.** Many species lack stipules, and in many where they do occur they fall off soon after the leaf has emerged from the bud. However, in peas and some other plants the stipules are large, permanent, and resemble the blades. The needles of pines could be considered as lacking blades.

The leaf blades of different species are indented to various degrees. If a leaf has no indentations it is said to be **entire.** If there are a few deep indentations, as in many oaks and maples, the leaves are **lobed** (the indentations between the lobes being the **sinuses).** The blades of some species are indented so deeply that the blade is divided into small leaflets, each of which is frequently mistaken for a complete leaf. Such leaves are **compound** (Fig. 46), in contrast with **simple** leaves that are either entire or lobed. In **pinnately compound** leaves, as of ashes, box elder, or black locusts, the leaflets are arranged along the sides of a single extension of the petiole (the **rachis),** which would be the midrib of the leaf if it were entire. In **palmately compound** leaves there is more than one main rachis (usually three or five), and they all emerge from the petiole at one point. The Virginia creeper or woodbine has palmately compound leaves with five leaflets. The leaves of poison ivy, however, are pinnately compound, even though they have only three leaflets.

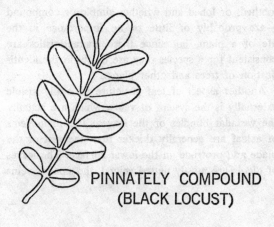

PINNATELY COMPOUND (BLACK LOCUST)

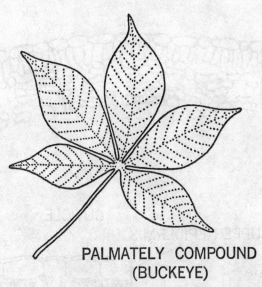

PALMATELY COMPOUND (BUCKEYE)

Fig. 46. Pinnately and palmately compound leaves.

Some pinnately compound leaves are twice compound, the leaflets being born on branches of the main rachis, and a few species such as the Kentucky coffee tree have thrice compound leaves with the leaflets born on branches of the branch rachises.

Compound leaves, in particular those twice or thrice pinnately compound, are frequently mistaken for branches bearing simple leaves. A simple way to determine whether a plant has simple or compound leaves is to look for axillary buds, since there is one present at the point where a leaf joins the stem but none where a leaflet is attached to a rachis. Also, a rachis does not have a terminal bud at its tip as does a stem. The margins of both leaves and leaflets may be either entire or toothed, as is characteristic of the species. The shapes of leaves—whether entire, toothed, or lobed and whether simple or compound —are probably of little or no importance in the life of a plant, but since these characteristics are consistent for a species they are useful in the identification of trees and other plants.

Another aspect of leaf structure generally visible externally is the system of **veins,** which are actually the vascular bundles of the leaves. The main veins of a leaf are generally thicker than the rest of the blade and protrude on the lower surface. The leaves of dicots are mostly netted veined, the smaller veins branching into somewhat irregular networks and the larger ones generally branching repeatedly. In contrast, the main veins of monocot leaves are usually parallel with each other, as in grass leaves, and the branch veins are rather small and inconspicuous. However, in both types of leaves no cell is very far from a vein. It should be emphasized that the vascular tissues (the xylem and phloem) of leaves are continuous with those of the stems, just as those of the stems are continuous with those of the roots. This provides a continuous pathway

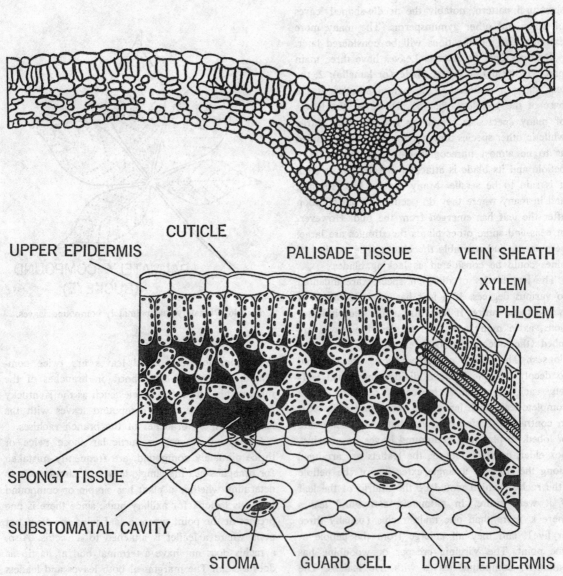

Fig. 47. Internal leaf structure. TOP: Section through a leaf blade showing the midrib and main vein. BOTTOM: Three-dimensional view of a small section of a leaf.

through which water, foods, and other substances may be carried quite rapidly from one part of a plant to another.

Internal Structure of Leaves. Although the details of internal leaf structure vary from species to species, consideration of only one of the more common patterns of internal structure will suffice here. The other structural patterns differ only in certain details. If a leaf is cut into thin sections perpendicular to the surface of the blade and observed under a microscope (Fig. 47), it will be found that both the upper and lower sides of the leaf are covered by an **epidermis** that is commonly only one cell thick and is covered with a cuticle. Under the upper epidermis is a row (or more rarely two rows) of elongated chlorenchyma cells oriented vertically in quite regular fashion. This is the **palisade layer.** Underneath it is a layer of more loosely and irregularly arranged chlorenchyma cells that are about isodiametric. This is the **spongy layer.** Together these two layers constitute the **mesophyll** of the leaf, the part of the leaf in which photosynthesis occurs. Extending through the mesophyll and located between the palisade and spongy layers are the veins, each with the xylem in the upper part of the vein and the phloem in the lower part. The veins are surrounded by parenchyma cells that constitute the bundle sheath. The larger veins may also have considerable sclerenchyma around them. The xylem is uppermost in veins because it is innermost in the vertical bundles of the stem, a simple matter of geometry. If a vein has been cut lengthwise in making the section ring or spiral, thickenings of the walls of the vessels are usually visible. Some veins of the leaf may end in a pore at the leaf margin.

Modified Leaves. Among the more common leaf modifications (Fig. 48) are the bud scales of woody plants. We have already mentioned that the flower parts—sepals, petals, stamens, and pistils—are all modified leaves and their derivation from leaves will be discussed more fully later. Flowers or clusters of flowers may also have somewhat less modified leaves located on the stem just beneath them, such leaves being known as **bracts.** Bracts are generally green and definitely leaflike in appearance, though they differ from the ordinary foliage leaves of the plant in size and shape. However, the white petal-like structures of the flowering dogwood and the red ones of the poinsettia are actually bracts. Flower heads of the members of the sunflower family have bracts around the base of the heads.

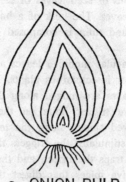

a. ONION BULB

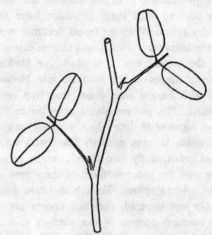

b. BLACK LOCUST THORNS

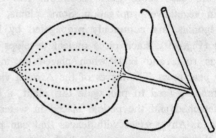

c. SMILAX TENDRILS

Fig. 48. Some modified leaves. a. The thick, fleshy scale leaves of an onion bulb. b. The thorns of black locust are modified stipules. c. The tendrils of this smilax are stipules.

The **spines** of barberry and some cacti are modified leaves, while in the black locust only the stipules are modified into spines. While some **tendrils** are modified stems, the tendrils of clematis are petioles of leaves and in peas and sweet peas some of the leaflets of the compound leaves are modified into tendrils. The **cotyledons** or seed leaves of seedling plants are modified leaves. The bulk of a **bulb** such as those of onion and tulip is composed of fleshy modified leaves.

Perhaps the most spectacular modified leaves are those of the carnivorous or insectivorous plants. In the Venus's-flytrap the leaves are hinged along the midrib and have sensitive trigger hairs on their upper surface and incurved spines along the leaf margins. When an insect stimulates the triggers the leaf suddenly snaps shut, traps the insect, and then apparently secretes digestive juices that digest the insect. Presumably this provides the plant at least with some protein food. In the sundew the small leaves are covered with long, glandular hairs that secrete sticky juices. When an insect becomes stuck on a leaf the hairs bend over it and secrete digestive fluids. In the water plant *Utricularia,* or bladderwort, some of the leaves are modified into bladderlike traps with hinged trap doors that trap small water animals. The pitcher plants have leaves that have grown together at their edges, forming vaselike structures which become partially filled with water. Insects that crawl or fly into these pitchers cannot get out because the inner surfaces of the leaves are lined with reflexed spines. Though digestive juices are probably not secreted, drowned insects are digested by bacteria present in the pitchers and the plant may absorb some of the digested food.

Vegetative Propagation by Means of Leaves. Leaves are used more than roots, but less than stems, in vegetative propagation. Some plants, notably begonias, are commonly propagated by leaf cuttings (Fig. 49). Each piece of leaf develops adventitious roots and buds when placed on moist sand or peat moss. The leaves of the African violet are commonly used to propagate this plant, a leaf being detached and the petiole placed in water or moist sand. Any species with leaves that can produce both adventitious roots and buds can be used for leaf cuttings. However, many species have leaves that can form adventitious roots but no buds, and

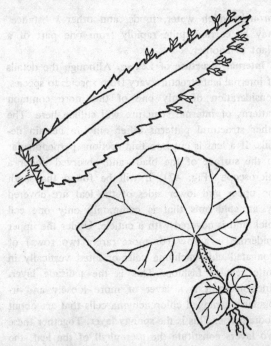

Fig. 49. Propagation by means of leaves. TOP: Plantlets developing on margin of a *Kalanchoe* leaf. BOTTOM: Plant developing from petiole of a *Begonia* leaf that has been detached and placed in moist sand. If *Begonia* leaf blades are cut into pieces and kept moist a plant will grow from each piece.

these cannot be used, while still other species have leaves that cannot even develop adventitious roots.

The leaves of *Bryophyllum* and *Kalanchoe* (air plants) naturally bear small plantlets in the notches of their leaf margins. These eventually fall to the ground and become established in the soil.

The long leaves of the walking fern propagate the plant by tip layering. When a tip touches the ground adventitious roots and buds develop and grow into a new plant, and its leaves in turn may then tip layer, the behavior that gave the plant its name. See Fig. 27.

Chapter 4

THE REPRODUCTIVE ORGANS OF FLOWERING PLANTS

Although vascular plants may reproduce asexually or be propagated by man by means of their roots, stems, or leaves, the basic type of reproduction is sexual and is carried on by special reproductive organs. The sexual reproduction of plants will be considered in some detail in Chapter 12, and in that chapter the reproductive processes of the vascular plants will be compared with those of lower plants and animals. Here we shall concern ourselves only with the more obvious aspects of the reproductive organs of the angiosperms, or the flowering plants. In other words, we shall be dealing here with flowers and the reproductive structures that develop from them—fruits and seeds.

FLOWERS

Flower Structure. As has already been pointed out, a flower consists of a specialized stem, with very short internodes and no apical meristem, that continues growing and bearing the flower parts (which are modified leaves). In a flower of the simplest and least modified type (Fig. 50) the lowest flower parts are the green and very leaflike **sepals** that make up the **calyx.** Just above the calyx is the **corolla,** composed of several **petals** that are typically white or some color other than green. The next set

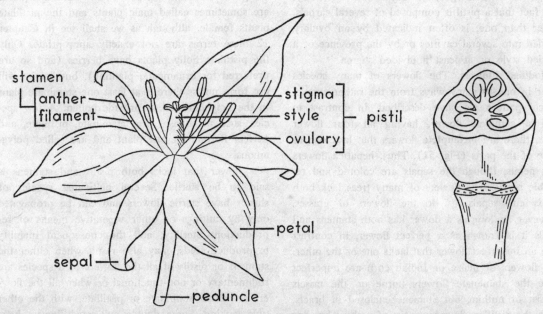

Fig. 50. A simple monocot flower. Except for the unusually short style, all the flower parts show well. The stigma is three-parted. RIGHT: A section across the ovulary (enlarged), showing the three cavities and ovules in them. Note the scars where the other flower parts were attached. This pistil is composed of three carpels.

of flower parts are the **stamens** (collectively the **androecium),** each consisting of a thin stalk (the **filament)** and the pollen-producing **anther.** Finally, the uppermost and most central flower parts are the **pistils** (or pistil), collectively called the **gynoecium.** A pistil consists of an expanded and hollow basal **ovulary** within which are borne the **ovules** that will develop into seeds, a stalk extending upward from the ovulary (the **style),** and an expanded tip of the style, the **stigma.** Each pistil is composed of one or more modified leaves, while each of the other flower parts is composed of a single modified leaf.

It is not too difficult to visualize sepals and petals as being modified leaves. If stamens are considered as leaves without blades and the filament as the leaf midrib, it is not too difficult to visualize them as modified leaves, either. Indeed, water lilies and some other plants have intergrading series of stamens and petals, some of the intermediate forms having both anthers and blades and looking quite leaflike. It is a little harder to visualize a pistil as one or more modified leaves, but if you will imagine that a leaf has folded around and grown together by its margins (as have the leaves of the pitcher plants) you can see how the hollow ovulary originated. In the case of a pistil composed of two or more modified leaves **(carpels)** the group of leaves can be considered to have merged, making a hollow ovulary. The fact that a pistil is composed of several carpels, rather than one, is often indicated by an ovulary divided into several cavities or by the presence of a divided style or at least a divided stigma.

Modified Flowers. The flowers of many species differ in one to many ways from the rather simple, typical pattern we have described. In contrast to such a **complete** flower, having all four flower parts, there are **incomplete** flowers that lack one or more of the parts (Fig. 51). Thus, hepatica flowers lack petals, though the sepals are colored and resemble petals. The flowers of many trees lack both petals and sepals, as do the flowers of grasses. However, as long as a flower has both stamens and pistils it is known as a **perfect** flower, in contrast with an **imperfect** flower that lacks one or the other. The flowers of maize or Indian corn are imperfect since the stamenate flowers borne on the tassels consist of nothing but stamens enclosed in bracts. Also, the pistillate flowers consist of nothing but one pistil each. The ovulary is the immature grain, and the silk is a long style with a stigma at the end.

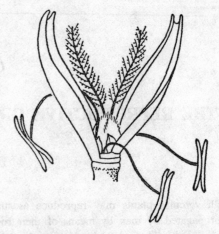

Fig. 51. A flower of grass, an incomplete flower. Grass flowers consist only of stamens and a pistil (or only one of these, as in corn). This pistil has two feathery stigmas on its ovulary. The two modified leaves on either side are neither petals nor sepals, but bracts.

An ear is a modified branch bearing numerous pistillate flowers enclosed in bracts (the husks).

When, as in corn, the stamenate and pistillate flowers are borne on one plant, the species is said to be **monoecious** (one household). Some hollies, willows, and other species, however, are **dioecious** (two households), the staminate and pistillate flowers being borne on separate plants. The staminate plants are sometimes called male plants and the pistillate plants female, although as we shall see in Chapter 12 these terms are not exactly appropriate. Only the pistillate holly plants have berries (and so are preferred for ornamental planting), but berries will not form unless there is at least one staminate plant in the neighborhood to provide pollen. A few species, such as squash, bear staminate, pistillate, and perfect flowers on one plant and are called polygamous.

A flower that lacks both pistils and stamens is said to be **sterile.** Several cultivated species of shrubs have sterile flowers and can be propagated only by cuttings or other vegetative means of reproduction. Sterility, and the consequent inability to produce seeds, may also result when either the stamens or pistils of all the flowers of a species are rudimentary or non-functional or when all the flowers are either staminate or pistillate, with the other type lacking. Some double cultivated flowers have had all stamens converted to petals, and so are sterile.

Other modifications of flowers result from the union of flower parts of the same set (**cohesion**) or of parts of different sets (**adnation**). Thus, the petals of the morning-glory are joined into a tubular, trumpet-shaped corolla, and another example of cohesion is found in the stamens of a bean flower, all but one of them being united to one another. Stamens are adnate to the petals of a good many species, and in cherry flowers both the petals and stamens are adnate to the sepals of the calyx. When the calyx and corolla are adnate and co-herent and form a tube that encloses (but is not attached to) the ovulary of the pistil, a flower is said to be **perigynous.** This is in contrast to the simple **hypogynous** type in which the calyx and corolla clearly are attached below the pistil and do not surround it (Fig. 52). An extreme type of adnation and cohesion is found in flowers, such as those of the apple tree, where the bases of the sepals, petals, and stamens are both coherent and adnate with one another and also adnate with the ovulary. In such an **epigynous** flower, the sepals, petals, and stamens appear to be attached to the top of the ovulary, though they are really attached below it as in any flower. The ovulary of an epigynous flower is inferior, in contrast with the superior ovularies of hypogynous and perigynous flowers. Fruits that develop from epigynous flowers are composed partially of the enlarged ovulary (as in any true fruit) and partially from the surrounding fused bases of the sepals, petals, and stamens.

In contrast to regular and radially symmetrical flowers, such as those of the strawberry or tulip, there are the irregular flowers of plants such as the sweet peas, snapdragons, or mints where either the calyx or corolla or both are modified so that they are asymmetrical in all but one plane, *i.e.,* they are bilaterally symmetrical.

The number of flower parts is characteristic of a species and is generally very constant for all flowers of the species. Indeed, numbers of parts is one of the general distinguishing characteristics between the dicots and monocots. Monocots have their flower parts in threes or sixes, while dicots generally have theirs in fours or fives or some multiple of these numbers. Thus, a tulip flower has three sepals (that look just like petals), three petals, six stamens, and a pistil composed of three carpels. Evening primrose flowers (dicot) have four sepals, four petals, eight stamens, and a pistil composed of four carpels, while the members of the rose and pea families have their flower parts in fives.

Inflorescences. The flowers of many species are not borne singly, but in clusters known as **inflores-cences** (Fig. 53). The individual flowers of an inflorescence are borne on branches or sub-branches of the main stem of the inflorescence, and the compactness of the inflorescence depends on how long the various internodes are and how highly branched the inflorescence is. The stalk of an individual flower, or the main stem of an inflorescence, is a **peduncle,** while the stalk of each individual flower in a cluster is a **pedicel.**

In the less compact types of inflorescences the

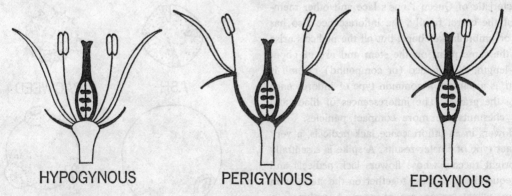

HYPOGYNOUS   PERIGYNOUS   EPIGYNOUS

Fig. 52. In a hypogynous flower (the simplest type) the ovulary is above the insertion of the other flower parts. In a perigynous type the other parts are united into a tube that surrounds, but is not attached to, the ovulary. In the epigynous type the fused bases of the other parts are attached to the ovulary, so they seem to come off from it.

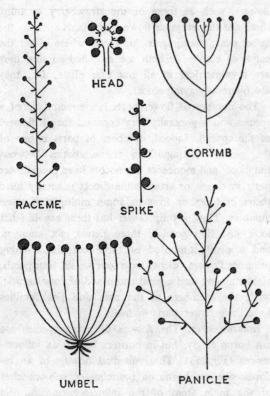

Fig. 53. Various types of inflorescences.

flowers have definite pedicels, sometimes rather long. A **raceme** is an inflorescence with a single main stem that continues apical growth and bears flowers along its sides, as in the hyacinth. A **corymb** is like a raceme except that the lower pedicels are longer than the upper ones, resulting in a rather flat cluster of flowers, as in the hawthorn. In an **umbel,** characteristic of Queen Anne's lace and other members of the carrot family, the inflorescence also has a flat or umbrellalike shape, but all the pedicels arise from the same point on the stem and are of about equal length. A branched (or compound) raceme is known as a **panicle,** a common type of inflorescence among the grasses. The inflorescences of lilacs and horse chestnuts are more compact panicles.

If flowers in an inflorescence lack pedicels, a very compact type of cluster results. A **spike** is essentially an upright raceme whose flowers lack pedicels and are frequently quite close together on the stem, as in mullein, wheat, or an ear of corn. A **catkin** is a spike that generally bears only pistillate or only staminate flowers, and the flowers usually lack petals and are in the axil of a scaly bract. Many trees such as birch, oak, willow, cottonwood, and walnut

have catkins. In the calla lily, jack-in-the-pulpit, and other plants of the same family, the small and numerous flowers are borne on a special type of upright, flashy spike known as a **spadix,** and this is more or less enveloped by a large bract known as the **spathe.** The white spathe of the calla lily is commonly mistaken for the corolla of a single flower. A **head** is a much shortened spike, as in clover or mimosa, where the inflorescence is globular. The most compact of all inflorescences are the heads of members of the sunflower or Composite family, including such common flowers as zinnias, marigolds, daisies, and dandelions. These inflorescences are commonly mistaken for a single flower with many petals, when actually each one is a cluster of flowers.

In the sunflower and daisy, two kinds of flowers are borne on the much-flattened and disc-shaped peduncle: the flowers of the outer circles have large strap-shaped corollas (the ray flowers) while those occupying the center of the disc have small, star-shaped corollas that are relatively inconspicuous. Each flower is epigynous, and the numerous "seeds" that develop from the flowers of a sunflower disc are actually fruits, each with a seed inside. Beneath the head of a composite there are generally numerous green bracts that may be mistaken for sepals if one is not aware that the "flower" is actually a cluster of flowers. In dandelions and chicory all of the flowers are ray flowers, so there is no central disc. Still other composites like the cocklebur have no ray flowers, and so their inflorescences are rather inconspicuous. How-

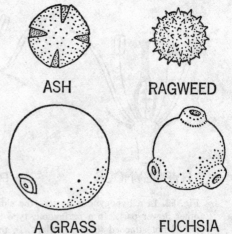

Fig. 54. Pollen grains of several different species of plants.

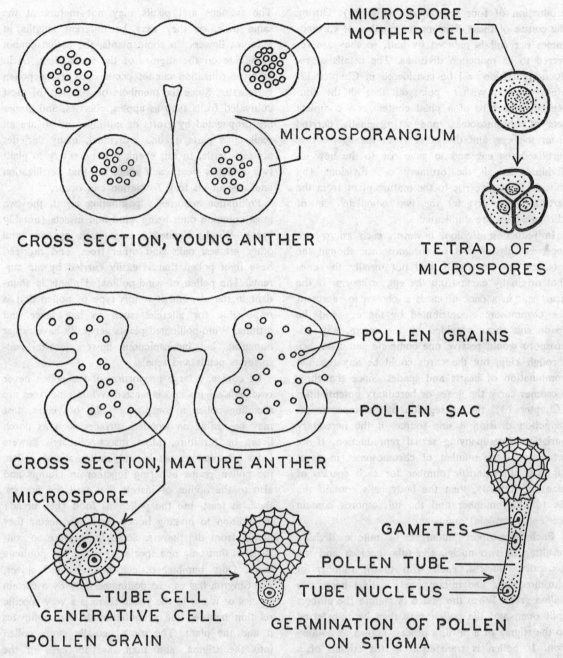

MICROSPORE
MOTHER CELL

MICROSPORANGIUM

CROSS SECTION, YOUNG ANTHER

TETRAD OF
MICROSPORES

POLLEN GRAINS

POLLEN SAC

CROSS SECTION, MATURE ANTHER

MICROSPORE

GAMETES

POLLEN TUBE

TUBE NUCLEUS

TUBE CELL
GENERATIVE CELL
POLLEN GRAIN

GERMINATION OF POLLEN
ON STIGMA

Fig. 55. The development of pollen in the anther of a stamen, and its germination after transfer to a stigma, shown diagrammatically.

ever, some inflorescences composed only of ray flowers, as in the thistle, are quite conspicuous.

Reproductive Processes of Flowers. Whatever the structural modifications of a flower may be, the reproductive processes of all types of flowers are quite similar. The stamens and pistils are the parts primarily involved. The anther of a stamen contains four **pollen sacs** within which the **pollen grains** (Fig. 54) are produced. At first numerous **pollen mother cells (or microspore mother cells)** are produced in the pollen sacs and each of these undergoes two successive cell divisions, resulting in the production of four microspores (Fig. 55). During the course of these divisions the number of chromosomes per cell is reduced by half, so they are referred to as reduction divisions. The details of reduction division will be considered in Chapter 12. Here we only wish to point out that all the non-reproductive cells of a plant contain two complete sets of chromosomes, one set originally derived from the **egg** and the other from the **sperm** that fertilized the egg and so gave rise to the new individual. In all the ordinary cell divisions (by mitosis) that give rise to the mature plant from the fertilized egg **(zygote)** the two complete sets of chromosomes are duplicated.

In reduction division, however, each microspore receives only one set of chromosomes, though the sets are not necessarily and not usually the ones that originally came from the egg or sperm. If the heart suit in a deck of cards is chosen to represent the chromosomes contributed by the egg and the spade suit those provided by the sperm, any microspore would receive one complete suit from ace through king, but the cards could be any possible combination of hearts and spades. Since the chromosomes carry the genes or hereditary potentialities (Chapter 13), this segregation of chromosomes at reduction division is one source of the hereditary variation accompanying sexual reproduction. If we let n equal the number of chromosomes in a set (n is a characteristic number for each species of plant or animal), then the body cells contain the $2n$ **(diploid)** number and the microspores contain the n (or **haploid**) number.

Each microspore undergoes a mitotic division resulting in two nuclei, the **tube nucleus** and the **generative nucleus,** each with the n number of chromosomes. The microspore has thus become a **pollen grain.** When the pollen is mature the anthers split open and the pollen may then be transferred to the stigma of a pistil, a process known as **pollination.** If pollen is transferred to the stigma of a flower on the same plant **self-pollination** has occurred, while transfer of pollen to the flower of another plant constitutes **cross-pollination.**

Some plants such as peas, beans, cotton, tobacco, wheat, and oats are normally self-pollinated, usually within one flower, but more species are cross-pollinated. Dioecious plants such as holly and willow can, of course, only be cross-pollinated. In other species there are various devices that make self-pollination, at least within one flower, impossible.

The stamens and pistils may not mature at the same time, or they may be different lengths in different flowers. In some plants, the pollen cannot germinate on the stigmas of the same plant, or in others fertilization cannot occur even if the pollen germinates. Since the members of a variety of most cultivated fruits such as apples, cherries, and grapes are propagated by grafts or cuttings and so are all really just parts of one individual, many varieties are self-sterile. In this event, it is necessary to plant two varieties near each other so that fertilization and subsequent fruit formation can occur.

Pollination requires a **pollinating agent,** the two most common ones being wind and insects (notably bees). Wind-pollinated species such as corn and other grasses, oaks and other trees, and ragweed have light pollen that is easily carried by air currents. The pollen of wind-pollinated plants is abundant in the air, and it is this type of pollen that is responsible for allergies such as hay fever and asthma. Wind-pollinated plants generally have rather numerous but inconspicuous flowers lacking conspicuous petals and sepals.

Of course, a large proportion of the pollen never reaches a stigma and is wasted. While pine trees are shedding pollen a conspicuous film of yellow dust may be visible on wooden surfaces such as porch floors or furniture. Most insect-pollinated flowers have conspicuous petals and rather sticky pollen, the pollen grains adhering together in clumps and also to the bodies of insects that visit the flowers. Bees, at least, use the pollen as food (bee bread) in addition to making honey from the nectar they obtain from the flowers. Some flowers are so constructed that only one species of insect can pollinate them. Only bumblebees can pollinate red clover, and Smyrna figs can be pollinated only by a certain species of wasp. In the yucca there is a very specific relation between the pronuba moth that pollinates it and the plant. The moth actually stuffs pollen into the stigma, and then lays its eggs in the ovulary. The moth larvae consume some of the developing seeds, but not all. The two species are thus quite interdependent for continued existence, or at least for reproduction.

Other less common means of pollination are by birds, particularly hummingbirds, or by water in some water plants. Of course, man may deliberately serve as a pollinating agent when he is crossing plants, either in genetic experiments or in an effort to breed new varieties of plants.

Once the pollen has been transferred from a stamen to the stigma of a pistil, pollination is complete. The pollen grain then begins forming an outgrowth known as the **pollen tube** that grows down the stigma and style and into the ovulary (Fig. 57). At this time the generative nucleus divides into two sperm, each containing the *n* number of chromosomes. One pistil may contain dozens or even hundreds of germinating pollen grains. When the end of a pollen tube reaches an ovule it breaks open and the two sperm are discharged into the ovule, following which **fertilization** occurs.

To understand just what happens we shall now have to go back in time and consider the development of an ovule (Fig. 56). In a very young ovule one central cell begins to get much larger than the surrounding ones. This is the megaspore mother cell, which then undergoes reduction division and produces four megaspores. One of these is much larger than the other three, and only this larger one survives and functions. It undergoes mitotic divisions resulting in eight nuclei (or another small number, depending on the species). Three nuclei migrate to one end of the enlarged megaspore, and walls are formed cutting them off as distinct cells. These three cells are called the **antipodals.** At the opposite end two cells are cut off (the **synergids**) and between them is the egg. The other two nuclei are located near the center and do not have walls around them. They are the **polar nuclei.** This entire structure derived from the megaspore is known as the **embryo sac,** and all the nuclei in it have the *n* chromosome number. The embryo sac occupies most of the volume of the ovule. Before fertilization the two polar nuclei fuse, resulting in the fusion nucleus with a 2*n* chromosome count.

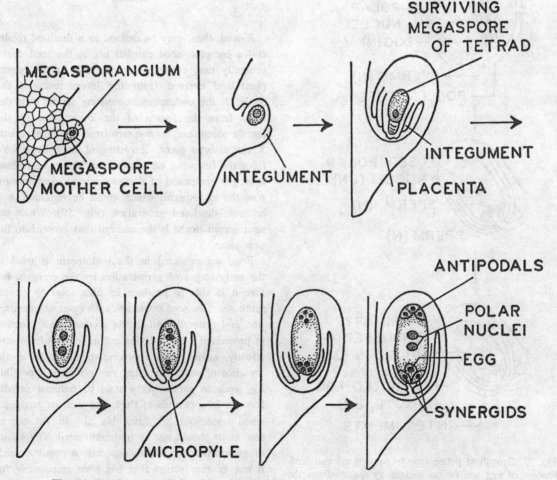

Fig. 56. Development of the embryo sac inside an ovule of a pistil, shown diagrammatically.

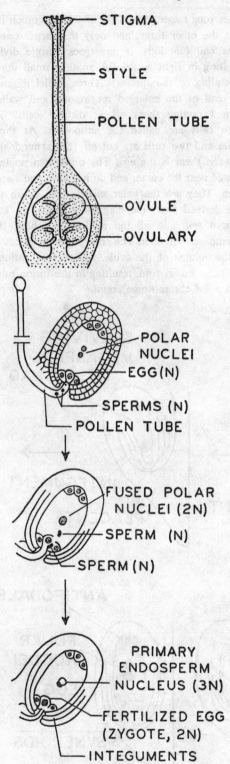

Fig. 57. Growth of pollen tube to embryo sac and fertilization of egg and fusion nucleus by sperms from the pollen tube.

After pollination has occurred and the pollen tube has reached the ovule the two sperm from the pollen enter the embryo sac (Fig. 57). One sperm fuses with the egg, resulting in a 2n zygote, or fertilized egg. This zygote then undergoes cell divisions resulting in an embryo plant (Fig. 58). The other sperm fuses with the fusion nucleus, resulting in the 3n primary endosperm nucleus. This then begins dividing by mitosis and so produces a multicellular 3n tissue surrounding the embryo. This tissue is the **endosperm**. In the meantime the synergids, the antipodals, and the tissues of the ovule inside its two-layered wall (the integuments) are generally disintegrating. The integuments remain and become **seed coats**. Inside the integuments is the endosperm, and inside it the embryo plant. Thus the ovule has been converted into a **seed**.

SEEDS

A seed, then, may be defined as a matured ovule, and a complete seed consists of: 1. the seed coats, generally two, consisting of 2n cells of the parent plant and derived from the integuments of the ovule; 2. the endosperm, consisting of 3n cells derived from the fusion of the 2n nucleus of the female parent and an n sperm from the male parent; 3. the embryo plant, consisting of 2n cells derived from the fusion of an n sperm and an n egg. Thus, a seed is composed of tissues from two generations, plus the endosperm, which could be regarded as a separate dead-end generation (Fig. 59). When the seed germinates it is the embryo that grows into the new plant.

Food accumulated in the endosperm is used in the early stages of germination by the embryo, before it is able to produce its own food by photosynthesis. The seed coats are split open at germination and generally fall to the ground, their function of protection against drying out or injury of the seed already having been performed. Since it is the presence of seed coats that really makes it possible for seeds to survive in a state of reduced activity through long periods of time, and then to germinate when conditions are favorable, the importance of seed coats should not be underestimated. The ability of gymnosperms and angiosperms to produce seeds is one of the factors that has been responsible for their success as land plants and for the fact that

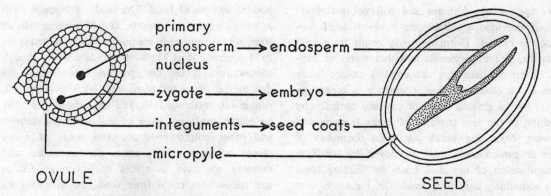

Fig. 58. Development of an ovule into a seed, shown diagrammatically. The embryo sac is in the last stage shown in Fig. 57 except that the antipodals and synergids have disintegrated.

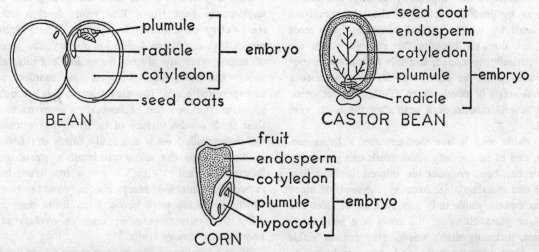

Fig. 59. Sections through the seeds of three species with large seeds. All the endosperm of the bean seed was used up before it matured, so only the embryo is found inside the seed coats. Note that the outer layer of a corn grain is fruit, not seed coats.

they represent the great bulk of the land plants of the earth.

The seeds of some plants such as peas and beans do not contain any endosperm, all of it having been used up before the seed matured. Such plants generally have thick cotyledons containing much accumulated food. Aside from this, seeds of all the many different species of seed plants do not differ much in basic structural composition, though they do differ greatly in size, shape, color, and structural details. The seeds of the coconut (a coconut without its fibrous husk) are among the largest, and at the other extreme are the microscopic seeds of orchids and a few other plants. Some seeds have special structures that aid in their dispersal. For example, seeds of milkweeds have tufts of hairs and pine seeds have wings, both contributing toward ease of transport by wind. Other species have seeds with spines or hooks that cling to the fur of animals or to clothing, and so may be carried some distance.

Although the seeds of many plants will germinate at any time that they are provided with suitable environmental conditions (water, oxygen, a suitable temperature, and sometimes light), the seeds of other species are dormant and will not germinate even if provided a suitable environment until dormancy is broken. Dormancy may result from seed coats that are impermeable to either water or oxygen or are so hard that the embryo cannot break them apart. Such seed coat dormancy is broken in nature by the gradual decay of the seed coats, or by cracking of the seed coats by alternate freezing and thawing. Man may break seed coat dormancy of farm or garden seeds, such as clover, by **scarification,** abrasion of the seed coats by shaking them with something like sharp sand. Or, the seeds may be placed briefly in strong acid that erodes the outer seed coat. The dormancy of the seeds of some species results from the presence of chemicals known as **germination inhibitors** in the seeds or the fruits. In nature this type of dormancy is broken by cold weather, which apparently destroys the inhibitors, or by gradual leaching of the inhibitors from the seeds by water. Still other species have seeds that are dormant because the embryo plants are only partially developed, and this type of dormancy is broken when the embryo finally has grown to a normal stage of development. This may not occur until several months or a year after the seeds were shed.

A **viable** seed is one that contains a living embryo, and of course only viable seeds can germinate. There has been considerable interest in how long seeds can remain viable. Actually, the seeds of many plants remain viable only one year, or at most two or three years. However, the seeds of a good many species, including many weeds, may remain viable for a dozen years or more, and some are known to live as long as eighty-five years. One reason that many weeds are so hard to eradicate is that many viable but dormant seeds are present in the soil, so they may appear even if all the weeds in an area were prevented from producing seeds for a year or more. There is no truth to the stories about six-thousand-year-old viable wheat seeds being found in Egyptian tombs, but viable water lotus seeds over two thousand years old have been found under the foundations of Japanese buildings of that age. The hard, thick coats of these large seeds enable the seeds to remain dormant over long periods and probably also aid in maintaining viability.

Man uses seeds in many ways other than in propagating farm and garden plants. Since seeds commonly contain much food in their endosperms or cotyledons, they provide one of our most im-

portant sources of food. The seeds of legumes such as beans, peas, and peanuts, and of grasses such as corn, wheat, and rice are particularly important as food sources. Most kinds of nuts are seeds, though commonly only the embryos are eaten. Seeds are the source of many plant oils such as corn oil, peanut oil, cottonseed oil, and linseed oil. The hairs on cotton seeds are used in making cotton fabrics and other cotton products. The seeds of many plants including celery, anise, dill, coriander, and caraway are used as spices or flavorings. Coffee and cocoa both come from seeds, as do castor oil and strychnine.

FRUITS

Only angiosperm plants have true fruits, and all angiosperms have fruits. True fruits develop from the ovulary of a pistil. Since seeds develop from ovules that are borne inside an ovulary, the seeds of angiosperms are always borne inside fruits, although the fruits may split open when mature and so expose the seeds. The exceptions to this are only apparent and not real. Strawberries seem to bear their seeds on the surface of the fruit, but actually the things called seeds are really small, dry fruits, produced from the many pistils of a strawberry flower. The fleshy "fruit" is not a true fruit, but rather the enlarged receptacle or stem end on which the pistils were borne. Such fruits that develop from structures other than an ovulary are known as **accessory fruits.**

Other angiosperms that may appear to have seeds not enclosed in fruits are the members of the sunflower and grass families, but in both families things we see are actually fruits rather than seeds. The so-called sunflower seeds are really fruits, and the real seed can be found by cutting open the fruit. Since each flower in the sunflower head is epigynous the fruit consists of both ovulary wall (the true fruit) and the fused bases of the other flower parts. In a grass such as corn the grain is actually a seed surrounded by ovulary wall, a true fruit. The seed has fused with the fruit at all points, and so is not loose inside the fruit like the seed of a sunflower. What we see when we look at the grain of corn, wheat, oats, or other grasses is not a seed coat, but rather a dry fruit. The apparent fruits of some gymnosperms, such as the "berries" of the juniper or red cedar and the "fruits" of the ginkgo are actually cones that are fleshy and more-or-less succulent, rather than hard and dry like most gymnosperm cones.

Fleshy Fruits. The fruits of angiosperms may be classified as dry or fleshy, each type including a number of different kinds of fruits based on structural features. The classification into fleshy and dry fruits is of little basic significance, one family of plants frequently containing species with both types of fruits, but it is a convenient classification. Few people other than botanists apply the term "fruit" to anything but fleshy fruits, and many people restrict the term even further to sweet fruits commonly used as desserts. Thus, tomatoes, cucumbers, squashes, eggplants, and green peppers are true fleshy fruits, but most people would class them as vegetables. Actually, they are both fruits and vegetables, the terms not being mutually exclusive. The term "vegetable" used in this sense is a culinary rather than a scientific word, and a vegetable may be a fruit, a seed (beans, peas), a flower (cauliflower, nasturtium), a stem (asparagus, broccoli), a bud (cabbage, Brussels sprouts), a leaf (spinach, endive), or a root (sweet potato, carrot).

The tissue of a fruit, whether derived from the ovulary or from tissues adhering to the ovulary, is referred to as the **pericarp.** Commonly three layers of pericarp can be distinguished: the outer **exocarp** (commonly called the skin), the **mesocarp** (the fleshy part of fleshy fruits), and the inner **endocarp** (which may be variously modified).

A fleshy fruit containing rather numerous seeds in a pulpy pericarp is a berry; examples are grapes, cranberries, currants, and tomatoes. (It is worth noting that fruits such as strawberries, raspberries, and blackberries are not really berries at all.) Citrus fruits, such as oranges, that have a thick leathery exocarp are a modified type of berry **(hesperidium).** The fruits of members of the squash family (watermelons, cucumber, cantaloupes, pumpkin) are called **pepos,** but are essentially large berries with thick rinds.

Another type of fleshy fruit is the **drupe,** or the stone fruits. Their endocarp is hard and stony and usually only contains a single seed. Examples are cherries, plums, and peaches. While the stones or pits of these fruits are commonly thought to be seeds it is necessary to crack them open to find the seeds.

Both berries and drupes are **simple fruits,** *i.e.,* they are derived from the ovulary of a single flower. Raspberries and blackberries are **aggregate fruits,** being composed of a cluster of small fruits derived from the numerous pistils of a single flower. The individual fruits of a blackberry are actually small drupes. Mulberries are also a cluster of small individual fruits, but in them each fruit was derived from the pistil of a different flower of a compact inflorescence. Such fruits are known as **multiple fruits,** in contrast with aggregate fruits. The pineapple also has multiple fruits.

Accessory fruits are those derived from tissues other than the ovulary, either entirely or partly. The fleshy strawberry is actually an enlarged receptacle, as has already been noted, while the true fruits of the strawberry are really small and dry (the "seeds"). Similarly, the true fruits of a fig are the hard and dry "seeds" derived from the pistils of numerous flowers of an inflorescence, while the fleshy fig is really the invaginated stem on which the flowers were borne. The outer layers of apples and pears are derived from the fused bases of the petals, sepals, and stamens, only the core being derived from the ovulary, and so they are largely accessory fruits. This type of fruit is known as a **pome.** Rose hips are also pomes, though few people think of them as being fruits.

Dry Fruits. The rather numerous types of dry fruits may be divided into two groups: those that split open when ripe **(dehiscent fruits)** and those that do not split open **(indehiscent fruits).** The indehiscent dry fruits generally contain only one seed and the pericarp has only a single obvious layer. **Achenes** are indehiscent fruits containing a single loose seed, as in the dandelion and sunflower and other members of that family and in the buttercup. The true fruits of the strawberry are achenes. Most **nuts** are essentially achenes, the term being restricted to fruits with large seeds and with woody seed coats or pericarp. In some nuts, however, the seed coats and pericarp are fused. The part of a walnut or pecan that we eat is the embryo plant, consisting mostly of the fleshy cotyledons. These nuts, along with almonds, are essentially dry drupes and differ somewhat from the true nuts such as those of the chestnut. The **grain** (or **caryopsis**) of such plants as wheat and corn differ from the achene in that the seeds are attached to the pericarp rather than being free from it.

The dehiscent dry fruits are referred to collectively as **pods.** Pods, such as those of beans, peas, and other members of the same family, that split open on two sides are called **legumes.** Pods that split open on only one side, as those of milkweeds and columbines, are called **follicles.** Both legumes and follicles are derived from simple pistils. If a pistil is composed of several carpels the resulting pod is called a **capsule,** as in mustard.

A few things commonly referred to as fruits are really not fruits at all. For example, rhubarb is really the fleshy petiole of the large leaves of this plant.

Significance of Fruits. Although the gymnosperms illustrate the fact that seed plants can flourish without fruits, the fruits of angiosperms play a number of different roles that probably contribute toward the success of many angiosperm species. Like certain seed characteristics, some fruit characteristics contribute to the dispersal of plants. Fleshy fruits are commonly eaten by birds and other animals and since the seeds are not digested they may be deposited on the ground at some distance from the

plant that produced them. The fruits of burdock, *Bidens,* and other plants have hooks or spines, similar to those on some seeds, and thus may be transported on the fur of animals. Maple fruits, like pine seeds, have wings that aid in dispersal by wind, while the fruits of dandelion have hairy parachutes that may carry the fruits and their contained seeds some distance. The ripe fruits of mistletoe stick to the feet of birds, and so may be carried to other trees. Some fruits, like the pods of witch hazel and touch-me-not, split open suddenly and so shoot their seeds some distance. Coconut fruits may float some distance through water, as is true of other species that may grow along shores.

Some fruits, like those of tomatoes, contain germination inhibitors. When the ripe fruits fall to the ground, the inhibitors keep the seeds from germinating until the following spring, thus keeping the young plants from developing at a time when they would be frozen before maturity.

Chapter 5

PLANT ACTIVITIES RELATED TO WATER

The importance of water in the lives of all living things cannot be overestimated. Life presumably began in the water. The great bulk of the plants and animals of the earth live in the oceans, and even the plants and animals of the deserts perish if they do not have an adequate supply of water. Protoplasm is mostly water, and the walls of living plant cells are impregnated with water. Water is essential in living things as a solvent, both because substances cannot be transported through plants unless they are in solution and because most substances will not react chemically with others unless they are in solution. In addition, water is an essential reactant in many very essential biochemical reactions such as photosynthesis and digestion. While a living organism can survive some time if deprived of food, death or inactivity soon results from a lack of water. Some seeds and spores can remain alive for long periods of time with a very low water content, and certain plants like the club moss known as the resurrection plant can become almost powdery dry and remain alive. But in all such cases life is going on at very low ebb, and active life will return only when water again becomes available.

DIFFUSION OF WATER

Since plant cells are completely surrounded by cell walls and the cytoplasmic membranes, it is apparent that water can pass through them into the cytoplasm and vacuoles of the cells; but the way in which water enters and leaves cells must be considered. While it seems likely that water may flow through the cell walls and perhaps even though the cytoplasm of plant cells, at least under certain conditions, entry of water into the vacuoles (and to a large extent into the walls and cytoplasm) is by diffusion.

Diffusion is quite a different process from the mass flow of water and should not be referred to as a flow. Diffusion results from the fact that the molecules of all substances are in constant motion (at least at any temperature above absolute zero, $-273°$ C.). If there is a net movement of molecules (or ions) of a substance from one region to another diffusion is taking place. This occurs when the substance has greater molecular activity (or greater diffusion pressure) in one place than in another adjacent place. When the molecular activity or the diffusion pressure of a substance is the same throughout a system, diffusion ceases because now just as many molecules are moving in one direction as another, so a dynamic equilibrium has become established.

There are three main factors that affect the diffusion pressure of a substance: 1. its **concentration,** that is, how many molecules there are per unit volume; 2. the **temperature,** since an increase in temperature results in an increase in the rate of molecular movement; 3. any outside or **imposed pressure** on the substance. An increase in the concentration of a substance, in the temperature, or in imposed pressure will increase the diffusion pressure of the substance.

Thus, while a plant is carrying on photosynthesis, oxygen is diffusing out of its leaves because it is being produced by photosynthesis and so is more concentrated in the leaf than in the outside air. At the same time carbon dioxide is diffusing into the leaf, because it is being used in photosynthesis and so is less concentrated than in the outside air (Fig. 60). Also, water vapor is diffusing out of the leaf because it is more concentrated in the leaf than in the air. Note that different substances may be diffusing in different directions at the same time and place, each one diffusing from where *it* has its highest diffusion pressure to where *it* has a lower diffusion pressure.

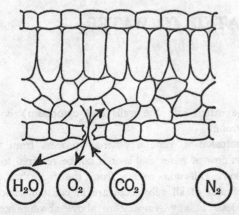

Fig. 60. When photosynthesis is occurring in a leaf, oxygen (produced by the process) is diffusing out through the stomate at the same time that carbon dioxide is diffusing in because its concentration in the leaf has been reduced by use in photosynthesis. Also, water vapor is diffusing out at the same time. Nitrogen is not diffusing in or out (since it is neither used nor produced and is in equilibrium). However, nitrogen molecules are moving in and out through the stomate at the same rate in each direction.

In a leaf that is the same temperature as the air (*i.e.*, a leaf in the shade) nitrogen is not diffusing into or out of the leaf because it is neither used or produced by the plant and has the same concentration inside as outside. The same number of nitrogen molecules are entering and leaving the leaf during any time period, and so there is a dynamic equilibrium and no diffusion. If, however, the leaf becomes exposed to the sun and so becomes warmer than the surrounding air, then nitrogen will diffuse out of the leaf for a while because the nitrogen molecules in the leaf are now moving faster than those outside, so their diffusion pressure is increased. There will soon be a new dynamic equilibrium, but when it is established the nitrogen in the leaf will be less concentrated than it is outside. Its diffusion pressure is the same inside as outside, however, because its molecules are moving faster; thus there are just as many molecules moving out as in. The faster movement makes up for the smaller number of molecules per unit volume.

In the examples we have just given there is no imposed pressure on any of the gases. Indeed, it is difficult to demonstrate the influence of imposed pressure on the diffusion pressure of a gas, since the pressure will compress the gas and so increase its concentration. However, liquids are not compressed to any degree by an imposed pressure and so their concentration does not increase, but their diffusion pressure does. For example, if water were placed in a watertight cylinder with a piston that could be subjected to pressure from a hydraulic press, the diffusion pressure of the water would be increased by the amount of pressure exerted on it. As we shall see, imposed pressure has an important influence on the diffusion pressure of water in plant cells.

First, however, let us consider how concentration may influence the diffusion pressure of liquid water. Whenever a substance is dissolved in water the solute particles are dispersed among the molecules of water (the solvent) and occupy space formerly occupied by water molecules. Thus, the water is less concentrated than it was. The more concentrated the solute, the less concentrated the water. In a 10% solution of sugar the sugar is more concentrated than in a 5% sugar solution, but the water is more concentrated in the 5% solution. Suppose we set up a rectangular battery jar with grooves at the middle on its bottom and sides into which a removable glass plate can be inserted and which will be watertight when in position (Fig. 61). With the plate in position let us place some 10% sugar solution on one side (A) and some pure water on the other side (B). Now let us remove the glass plate separating the two. Since sugar is more concentrated in A than B it will now begin diffusing from A to B and will continue to do so until it has the same concentration throughout, *i.e.*, 5% in this case. At the same time water will be diffusing from B to A, because it is more concentrated in B than in A. At equilibrium, it too will be the same concentration throughout. In this example the diffusion pressures of both water and sugar are being influenced only by differences in concentration, since there is no temperature difference and no differential imposed pressure.

Note, however, that diffusion did not begin until the glass plate was removed. The plate served as an **impermeable membrane,** through which neither water nor sugar molecules could pass. Now let us repeat the demonstration substituting a **permeable**

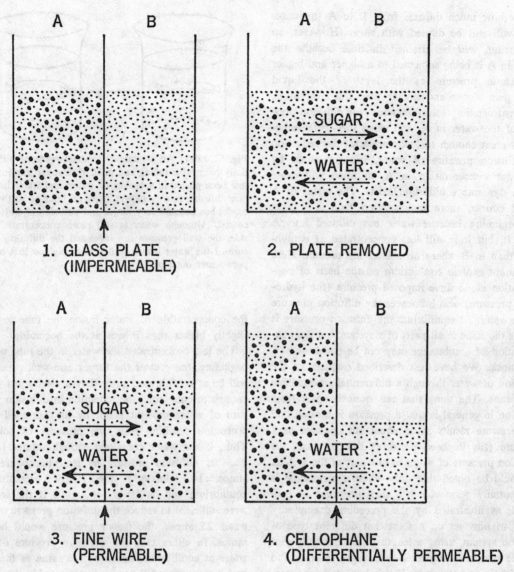

Fig. 61.

membrane (*e.g.*, a piece of fine screening) for the impermeable glass plate. The permeable membrane permits all kinds of molecules to pass through it and so diffusion proceeds just as if no membrane were present, except that attainment of the equilibrium may be somewhat delayed because some of the molecules bump into the wires of the screen and bounce back, instead of proceeding to the other side.

Now let us insert the third possible type of membrane between the two sides of the jar—a **differentially permeable** membrane through which wa-

ter molecules but not the larger sugar molecules can pass because the latter are too big to get through the very small pores in the membrane. (The differentially permeable membrane might be a sheet of cellophane from which the waterproofing substance has been removed by dissolving it out with alcohol.) We now have quite a different situation than before. Water can still diffuse from B to A and will do so, but sugar cannot diffuse from A to B. The result will be that the level of liquid in A will rise while it is falling in B. Water will never become as concentrated in A as in B because no matter how much diffuses from B to A the water in A will still be diluted with sugar. However, an equilibrium *will* be attained in time because the water in A is being subjected to a higher and higher hydrostatic pressure as the level of the liquid in A gets higher and higher than the level in B. This hydrostatic pressure increases the diffusion pressure of the water in A, and when this increase becomes great enough to just counteract the reduction of diffusion pressure of water, brought about by its lower concentration in A, diffusion will cease and a dynamic equilibrium exists. The water in A is, of course, more concentrated than it was at the beginning because water has diffused into A from B, but it is still less concentrated at equilibrium than in B where no solutes are present. Here we cannot explain equilibrium on the basis of concentration alone since imposed pressure (the hydrostatic pressure) also influences the diffusion pressure of the water. At equilibrium the diffusion pressure is always the same in all parts of a system, but the concentration of a substance may not be the same.

Osmosis. We have just described osmosis, or the diffusion of water through a differentially permeable membrane. The thing that sets osmosis apart from diffusion in general is that a pressure is created and this pressure results in an equal but opposite back pressure (the imposed pressure) that increases the diffusion pressure of water in one part of the system. It should be noted that only when a membrane is differentially permeable will such pressures be produced, as illustrated by the preceding examples.

Let us now set up a somewhat different type of osmotic system, using a length of differentially permeable plastic tubing filled with sugar solution and tied tightly at both ends. If this is placed in a container of water the water will diffuse into it but the sugar cannot diffuse out (Fig. 62). The result is an increase in the amount of water inside the tube, which thus becomes inflated (**turgid**) as a tire becomes inflated when air is pumped into it. This pressure is known as **turgor pressure,** and the back pressure of the membrane against the contained solution (the imposed pressure) is known as **wall pressure.** Since there was not much room for added water in the tube turgor pressure and wall pressure increase quite rapidly and so the diffusion pressure of water becomes the same inside as out after only a little has diffused in. (The tubing cannot stretch much.) Here the imposed pressure plays a very important part in increasing the diffusion pressure in the tube, since

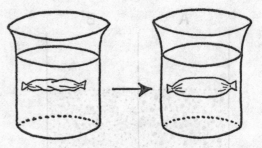

Fig. 62. LEFT: An osmotic cell model made of differentially permeable tubing and filled with sugar solution has just been placed in a beaker of water, and water has begun diffusing into it. RIGHT: After an hour or so the cell model has become very turgid and diffusion of water has ceased. Although water is still more concentrated outside, the wall pressure has increased the diffusion pressure of the water inside until it is as high as that of the pure water outside.

the concentration of water inside the tube is only slightly higher than it was at the beginning.

The less concentrated the water in the tube at the beginning, the greater the turgor and wall pressures will be at equilibrium, since they must be just great enough to counteract the reduction in diffusion pressure of water brought about by its decreased concentration resulting from the presence of solutes. Thus, if the water contained enough solutes to reduce its diffusion pressure by ten atmospheres (1 atmos.$=14.7$ lb./in.^{2}) the turgor pressure at equilibrium will be 10 atmos., while if the solutes were sufficient to reduce the diffusion pressure of the water 25 atmos. the turgor pressure would be 25 atmos. In either case the diffusion pressure of the water at equilibrium would be the same as the diffusion pressure of the pure water outside the tube, even though it is much less concentrated.

If the tube is placed in a solution rather than pure water, the diffusion pressure of the water inside will be the same as outside at equilibrium and the turgor pressure will be less than the possible maximum in pure water. For example, if the diffusion pressure of the water inside the tube were reduced by 25 atmos. and that outside by 18 atmos. the turgor pressure at equilibrium would be only 7 atmos. In all these examples we have been ignoring the slight changes in concentration of water inside and out resulting from diffusion.

If the water in the tube had a reduction in diffusion pressure of 10 atmos. and that outside of 18 atmos. the water would diffuse out rather than in because the diffusion pressure would be greater inside. At equilibrium the concentration of water as well as its diffusion pressure would be the same inside and out, since there would be no wall pressure. The less concentrated the outside water, the more water would diffuse out of the tube before an equilibrium was attained.

Osmosis in Plant Cells. As far as osmosis is concerned, the tubes we have been describing are quite comparable with (though much simpler than) plant cells. The cytoplasmic membranes are differentially permeable (though they permit some solutes as well as water to diffuse through them) and the vacuoles contain water diluted by the presence of solutes, corresponding with the solution inside the tube. The cytoplasm, however, unlike the plastic tubing, is quite stretchable and would balloon out as water diffused in if it were not for the restraining cell wall. The cell walls thus play a role comparable with that of a tire casing, which prevents the ballooning of the inner tube. The cytoplasmic membranes are a mosaic of proteins and fatty substances. Water and water-soluble substances enter through pores in the protein portions, and so the membranes are impermeable to water-soluble molecules too large to get through the pores. However, fats and fat-soluble substances enter through the fatty portions by dissolving in them and so the more soluble a substance is in fats the easier it can cross the membrane, regardless of the size of the molecules up to a point.

The diffusion pressure of water in the soil is

generally higher than in the root cells, and so it diffuses into the root hairs and epidermal cells from the soil. As long as there is any capillary water present in most soils its diffusion pressure is only about 1 atmos. less than that of pure water, while in the root cells it is commonly around 5 atmos. less. In the cells of the stems and leaves it may be from 8 to 30 or more atmos. less, the diffusion pressure of water in the cells of trees generally being lower than in those of herbaceous plants. Water will diffuse from one cell to another whenever there is a difference between the diffusion pressure of the water in them, and there are usually gradients of diffusion pressures across the tissues of a root, stem, or leaf. It should be noted that the diffusion pressure of the water in any turgid plant cell is determined by at least two factors: the concentration of the water and wall pressure. Evidence secured in recent years indicates that the rate at which water enters plant cells by osmosis is sometimes more rapid than could be accounted for by simple diffusion, and it seems likely that some of the water is entering by mass flow rather than diffusion. However, this does not alter any of the points made in the preceding discussion of osmosis.

Turgor Pressure of Plant Cells. Perhaps the most important aspect of osmosis in plant cells is the turgor pressure created. When cells lose their turgor pressure the plant wilts; herbaceous plants, at least, depend on turgor pressure for support. Cells that are not turgid are operating at a low level, and many important processes such as photosynthesis are going on at a much reduced rate. Growth cannot occur when cells lack turgor, although an enlarging cell generally has a lower turgor pressure than it did just before it started enlarging simply because of the increasing size of the wall. This results in a decrease of diffusion pressure of water in the cell and so in diffusion of more water into the cell. There is generally a daily fluctuation in turgor pressure of cells, the low point usually being in the afternoon and the high point late in the night or early in the morning. Leaf cells frequently have turgor pressures of from 5 to 25 atmos. or so. A turgor pressure of 20 atmos. would be 294 lb./in.^{2}, a much higher pressure than is ever found in a tire or a steam boiler, yet cell walls are generally strong enough to resist such high pressures. However, at times such high turgor pressures develop that cells actually explode or have "blow-outs." One condition under which this may occur is when a pollen

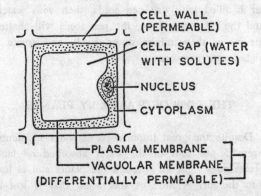

Fig. 63. A plant cell contains a solution (the cell sap) and has differentially permeable membranes (the plasma and vacuolar membranes, particularly the latter). The cell wall is permeable, but resists turgor pressure and exerts an imposed pressure (wall pressure) on the water in the cell.

grain is in contact with a drop of rain water and so develops its maximum turgor pressure. This is one reason that extended rains while apple trees are blooming may reduce the apple crop, many of the pollen grains exploding. Also, the cells of ocean or salt marsh plants (which have a high solute content and thus a low water concentration) may explode when they are placed in fresh water.

Plasmolysis. When cells are placed in a solution in which water has a lower diffusion pressure than the water in the vacuoles, water will diffuse out of the cells. As it does so the elastic cytoplasm shrinks away from the cell walls (Fig. 64). This is known as **plasmolysis.** If plasmolysis has not lasted too long

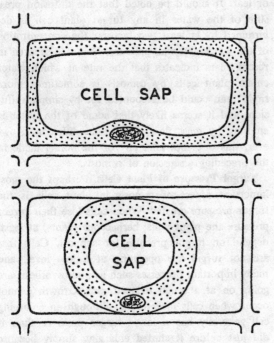

CELL SAP

CELL SAP

Fig. 64. If the water surrounding a cell has a lower diffusion pressure than the water in the cell sap it will diffuse out. Turgor pressure becomes zero and the elastic cytoplasm contracts from the wall, first only at a few places (top). As more water leaves the cell sap the cytoplasm may shrink into a sphere (bottom). In both sketches plasmolysis has thus occurred. If there were only a slightly lower diffusion pressure outside, plasmolysis might not proceed beyond the initial stage shown in the upper drawing. The external solution fills the space between the plasma membrane and the cell wall.

the cell may regain turgor when placed in a solution with water of a higher diffusion pressure than that in the cell, but extended plasmolysis may result in the death of the cell by desiccation. Plasmolysis

occurs when too much fertilizer is applied to plants. The plants then dry up and die **(burning).** Land or fresh-water plants are plasmolyzed when supplied with ocean or salt-marsh water. In some regions irrigated for many years the salt content of the soil has become so high that ordinary crop plants or citrus trees can no longer be raised, since they are plasmolyzed and killed. This is also what happens when a shipwrecked person attempts to drink the salty ocean water, water actually diffusing out of his intestinal walls rather than in. Only plants that have a very high concentration of solutes (and so a very low concentration of water) in their cells can grow in oceans, salt marshes, or alkaline soils.

Imbibition. Not all the water that enters a plant tissue enters by osmosis. Dry plant tissues such as seeds absorb much water by **imbibition.** Imbibition occurs when a substance such as the cellulose in cell walls has an affinity for water. After the water diffuses into the walls the cellulose holds the water molecules with great force and reduces the rate of movement of the molecules, thus reducing their diffusion pressure greatly, the lost energy of the molecules appearing as heat. Then more water diffuses in. The water molecules held by the cellulose cause it to swell. The swelling of seeds when soaked in water or of rice while it is being cooked is a result of imbibition. The swelling of doors and windows is also imbibition, the water coming from the water vapor of the moist atmosphere. If imbibition occurs in a restricted volume that resists swelling, very high turgor pressures may result—up to 1000 atmos. or so, in contrast with turgor pressures of less than 50 atmos. as a result of osmosis. Thus, if a glass jar is filled with dry pea seeds, then with water, and the lid screwed on, the pea seeds will shatter the jar as they imbibe water, swell, and create a high turgor pressure.

THE LOSS OF WATER BY PLANTS

Despite the great importance of water in plants, over 95 per cent of the water absorbed by land plants goes right on through the plants and is lost into the air. A little of this water that is lost is forced out of the open ends of veins at leaf margins as drops of liquid water, a process known as **guttation,** but most of it evaporates from the leaves, stems, and other above-ground parts of the plant into the air as water vapor. This loss of water vapor from plants is known as **transpiration.** Transpiration really consists of two steps: 1. the evaporation of water from the wet cell walls into the intercellular spaces; 2. diffusion of this water vapor through stomates, lenticels, or the cuticle into the outside air. Most of the water that transpires from leaves or herbaceous stems passes through stomates, only a small percentage diffusing through the cuticle. Transpiration through lenticels is restricted, of course, to stems with bark.

Magnitude of Transpiration. The quantity of water lost by transpiration is very great. For example, a fifty-foot high red maple tree may lose six gallons of water a day, or almost a thousand gallons during a season, while an acre of corn loses around 300,000 gallons during a growing season. The rate of transpiration is not steady, but varies greatly during a day and from day to day. Generally the rate is very low at night, increases steadily through the morning to a high point by early afternoon, and then decreases again. Transpiration is more rapid in dry regions than in humid ones, and more rapid when the temperature is high than when it is low.

The rate of transpiration, as long as the stomates are open, is determined largely by the difference in the diffusion pressure of the water in the intercellular spaces and in the air, *i.e.*, the steepness of the diffusion pressure gradient. There will obviously be a more rapid rate when the air is dry than when it holds much water vapor. The increase in rate of transpiration with temperature is very marked, being three or four times as fast at 80° F. as at 50° F. The reason for this is that as air gets warmer it can hold more water vapor, and in the intercellular spaces the air is always almost saturated with water vapor. Thus, with an increase in temperature the concentration of water vapor in the leaf increases greatly. In the outside air, however, the concentration of water vapor remains about the same, even though the water-holding capacity of the air increases. The result is a much steeper diffusion pressure gradient than existed at the lower temperature and so an increased rate of transpiration.

Stomates and Transpiration. When the guard cells of a stomate are turgid the stomate is open and when the guard cells lose their turgor pressure they collapse against one another and the stomate closes. The walls of the guard cells adjacent to the stomatal pore are thicker than at other places and do not stretch as much as the thinner walls when the guard cells are turgid. This results in an inward cupping

(similar to the inward cupping of an inflated inner tube where there is a patch) that separates the guard cells and makes the stomate open.

When a plant becomes deficient in water, as in the afternoon of a hot day when the loss of water by transpiration exceeds the absorption of water, the guard cells lose their turgor and the stomates close. If it were not for this stomatal closing, transpiration would continue at a rapid rate and the plant might die from lack of water. However, the stomates do not close in order to save water—they close because the guard cells lose their turgor pressure. The closing of stomates during mid-afternoon of a hot day is responsible for the decreased rate of transpiration at that time, even though the diffusion pressure gradient is very steep.

Stomates are generally closed at night and open in the morning when it becomes light, not "in order to enable carbon dioxide to enter so that photosynthesis can occur" but because light causes an increase in the turgor pressure of the guard cells. The sequence of events is apparently something like this: 1. light causes photosynthesis to begin in the guard cells; 2. this uses up CO_2 and therefore the carbonic acid formed when CO_2 dissolves in water; 3. the reduced acidity of the guard cells favors the conversion of starch to sugar; 4. since sugar is soluble and starch is not, this results in a decrease in water concentration in the guard cells; 5. the reduced diffusion pressure of the water in the guard cells results in the diffusion of more water into them, and so an increase in their turgor pressure. At night the reverse series of events causes stomatal closure.

The number and size of stomates is a species characteristic. Many plants have stomates only on the lower epidermis of the leaves, and species that do have stomates on both surfaces generally have more on the lower one. However, grasses generally have more on the upper epidermis. The number per square centimeter varies greatly from species to species, the wandering Jew plant having only 1400 while some oaks have 100,000. However, plants with fewer stomata generally have larger ones. The diffusive capacity of the stomata of all plants seems to be great enough so that when the stomates are open it is the steepness of the vapor pressure gradient that determines the rate of transpiration rather than any limitation of transpiration imposed by the stomates. Small pores like stomates are very efficient diffusion pathways. Even though the stomatal openings of a leaf occupy only about 1 to 3 per cent of the leaf area, they permit about 80 per cent of the diffusion that would occur if the leaf had no cuticle and diffusion occurred from its entire surface.

The Significance of Transpiration. Botanists have for a long time been interested in the question of whether or not transpiration plays any essential or useful role in the life of plants, and for a long time the consensus was that it does not. The most obvious possible role is the cooling effect on leaves and other organs, since evaporation does have such a cooling effect. However, transpiration is only one of several means of heat loss from leaves and it was believed that reradiation and convection were more important means of heat loss and that there would be no heat injury to leaves in direct sunlight even if there were no transpiration. In recent years more extensive and critical measurements of heat loss by transpiration have made it clear that under conditions of high temperature and intense sunlight transpiration is actually an important factor in preventing heat injury of leaves, particularly in certain species.

It has been stated that transpiration is essential for bringing water to the tops of plants, and it is largely responsible for the rate of water flow up the stems of plants. However, as we shall see soon, both methods of water transport in the xylem of plants could go on even though there were no transpiration. Finally, it has been suggested that transpiration might be necessary for the absorption and transport to the tops of plants of enough of the essential mineral salts. Some experiments do indicate that more mineral salts reach the tops of plants when transpiration is rapid than when it is slow, but apparently enough could be delivered even without any transpiration.

In summary, the cooling effect of transpiration is important under certain conditions and it may also contribute toward the absorption and transport of mineral salts. However, it is not necessary to invoke such roles of transpiration in an explanation as to why it occurs. The stomates are really essential as channels of diffusion of carbon dioxide and oxygen, the gases exchanged with the atmosphere during respiration and photosynthesis, and since the stomata are present there is nothing to prevent the diffusion of water vapor through them. This loss of water vapor at a region where gas exchanges are occurring is comparable with the loss of water through the lungs of an animal, not with perspiration which is really excretion by the sweat glands. Transpiration

is often of considerable significance in a negative way, since plants would probably thrive better if they did not lose most of the water they absorb and would not deplete the soil of water so rapidly. If plants die after transplanting it is usually because they dry out because transpiration exceeds absorption from the injured root system. This is why leaves are sometimes removed from plants when they are transplanted and why paper caps are placed over young plants transplanted into a garden or field.

Permanent and Temporary Wilting. If a plant wilts because of a deficiency of water in the soil the wilting is permanent. This does not mean that the plant will never recover from wilting, but it does mean that it will recover only after more water has been added to the soil, provided it has not died. Temporary wilting occurs even though the soil contains an abundance of water when the rate of transpiration is faster than the rate of water absorption. Temporary wilting commonly occurs every afternoon during hot summer weather, and the plants recover after the stomates have closed and after the temperature falls at night. Adding water to the soil will not make plants recover from temporary wilting. The only way to make a plant recover from temporary wilting is to reduce the rate of transpiration. A wilted plant, whether permanently or temporarily wilted, is carrying on its life processes slowly and is not growing. Photosynthesis is particularly hampered by a lack of carbon dioxide because of the closed stomates. If there were no transpiration, temporary wilting would never occur and there would be much less permanent wilting because the soil water would not be depleted so fast.

THE TRANSLOCATION OF WATER

Diffusion of liquid water or of solutes is quite slow compared with the rapid diffusion of water vapor and other gases, and if land plants were dependent on diffusion of water from cell to cell from the roots to their tops they could grow to a height of only a few inches. Actually, the bulk of the water transported from the roots to the stems, leaves, and flowers of a plant flows through the tracheids or vessels of the xylem at a much more rapid rate than it could diffuse from cell to cell. Although it is known that water flows through the xylem, identification of the forces responsible for this upward flow of water has been rather elusive.

It is known that several of the forces that are commonly supposed to be involved are quite inadequate to account for the rise of water to the top of even a small tree, let alone the tallest trees which approach 400 feet in height. The diameters of the tracheids are such that capillary action could account for the rise of water to a height of only four or five feet, while in tubes the size of vessels capillary rise would be only to a height of five inches or so. Air pressure could not possibly act, since it can support a column of water only about 32 feet high, and even this demands that there be a vacuum at the top and that the bottom of the tube and the water in which it is immersed be exposed to the air, which is not true of the ends of the xylem in the roots. Several botanists have proposed theories of water rise involving the action of the living parenchyma cells in the xylem, but since water will continue to flow through the xylem even if these cells have been killed none of these theories is regarded as acceptable. Only two of the many theories proposed seem to fit the facts and to represent actual mechanisms of water translocation: root pressure and shoot tension.

Root Pressure. To state the matter as simply as possible, root pressure results from the continued diffusion of water from the soil, through the tissues of the root, and into the xylem of the root. As more water enters the xylem it pushes up that which is already present. The operation of this mechanism can be demonstrated by cutting off the top of a plant and attaching a piece of glass tubing to the stump with rubber tubing. The water will be seen to rise in the glass tubing. By substituting a manometer for the glass tubing it is possible to measure the pressure developed.

Despite the fact that root pressure does occur in some plants part of the time, it is apparently not the way in which water rises in most plants most of the time. For one thing, root pressure has never been found in many species, and in the species where it has been found it occurs only under conditions of high soil moisture and high humidity—the very conditions under which flow of water through the xylem is the slowest. Furthermore, root pressures of more than 2 atmos. have never been measured in intact root systems. This would only be able to explain the rise of water to a height of 64 feet. Considering the fact that half the available energy is used in overcoming the friction of the water against the walls of small tubes such as tracheids or

vessels, root pressure would, at best, explain the rise of water to a height of only 32 feet or so. In addition, root pressure moves water quite slowly and cannot account for the rates at which water is known to flow through the xylem.

It appears, then, that root pressure does operate, but that it can explain the rise of water only in some plants part of the time.

Shoot Tension. The method of water transport that appears to operate in most plants most of the time is shoot tension, also referred to as the **cohesion mechanism** or the **transpiration-cohesion-tension mechanism.** In this mechanism the water is actually *pulled* up the xylem, not pushed up as it is by root pressure (or as it would be by air pressure if it could operate). Shoot tension is possible because water in small, airtight tubes has a great cohesive strength. A pull of over 30 atmos. (441 lb./in.^{2}) is required to pull such a column of water apart, so the water has greater cohesive strength than a steel wire of the same diameter. However, the water must have little or no air dissolved in it since the air comes out of solution and forms bubbles that break the column as the water is subjected to tension (negative pressure).

Shoot tension has been demonstrated by attaching a branch of a tree to a glass tube over 32 feet long, and it has been shown that the branch can raise the water considerably higher than atmospheric pressure could pull it. If a shorter piece of glass tubing filled with water is immersed in mercury, it has been shown that a branch can pull the mercury higher than the 76 cm. to which air pressure could push it, the mercury having risen to 100 cm. or more in some such experiments. The same thing can be done with a completely mechanical system in which a porous clay tube from which water can evaporate is substituted for the branch.

Thus, shoot tension is seen to be a force that can operate. However, three questions remain: 1. what are the motive forces that pull the water up?; 2. are the forces great enough to pull water to the tops of the tallest trees?; 3. is the water in the xylem actually under a tension?

It appears clear that the motive force is the diffusion of water from the xylem into the cells of the leaves and stems of a plant. As water diffuses out of the xylem, the water behind it is pulled up because of the great cohesive strength of water. The main factor that reduced the diffusion pressure of water in stem and leaf cells, and so causes water to diffuse into them from the vessels or tracheids of the xylem, is the loss of water by transpiration. The rate of transpiration is largely responsible for the rate of water flow through the xylem. However, anything that uses water in the cells (such as photosynthesis or digestion) will result in a decrease in the diffusion pressure of water in the cell and so in diffusion of water into it from another cell or directly from the xylem. Thus, transpiration is not essential for creating the reduction in diffusion pressure of water that provides the motive force for pulling water upward.

If this motive force is to be great enough to raise water to the tops of the tallest trees it must be able to exert a pull (tension or negative pressure) of around 25 atmos., considering that half the energy would be used in overcoming friction. (25 atmos. tension ×16 ft. effective rise/atmos.=400 ft.) This means that the diffusion pressure of the water in the leaf cells would have to be at least 25 atmos. less than that of the water in the soil. This has been shown to be the case.

Thus shoot tension appears to be theoretically possible and it remains to be demonstrated that the water in the xylem is actually under a tension rather than under pressure. This has been done by several different experimental techniques. One of the simplest is to remove carefully the tissues external to the xylem of a herbaceous stem and then to observe a vessel under a microscope. If the vessel is punctured with a fine needle the water in it is seen to snap apart, showing that it was under tension. If it had been under pressure, the water would have oozed or squirted out. Also, it has been found by the use of delicate measuring instruments that the diameter of a tree trunk is slightly less during the day (when water flow is fastest and the tension is the greatest) than during the night. This would be expected if the water is under tension, since the diameter of each vessel would decrease slightly, but if water were under pressure the vessels would be expected to increase in diameter slightly.

Shoot tension thus appears to be both theoretically feasible and the actual method of water rise in most plants most of the time. Though a few botanists still have doubts regarding it, most feel that it provides the only satisfactory explanation of water rise to the tops of the tallest plants. One point that has created opposition to the theory is that water still continues to rise in a tree even if the water columns in the vessels or tracheids have been broken. How-ever, there is now evidence that in such cases the water may be pulled through the walls of these cells rather than through their cavities.

THE ABSORPTION OF WATER

In our reverse consideration of the water economy of plants we now come to the first step in point of time: the absorption of water from the soil. It should soon become evident that an understanding of water absorption depends on some understanding of the loss and translocation of water. Roots are, of course, the primary organs of water absorption but plants may absorb some water through their leaves or stems under certain conditions. For example, submerged water plants generally absorb water through any part of their surface as do epiphytes such as Spanish moss that have no roots in the soil and grow on trees, though they secure neither food nor water from the tree. Two types of water absorption by roots have been identified: active and passive.

Active Absorption. Active absorption of water requires living root cells and a diffusion pressure gradient between the soil water and the water in the xylem of the roots. Water diffuses by osmosis from the soil, through the tissues of the root, and into the xylem. When water is being absorbed actively root pressure results, the two processes being linked. Guttation occurs only when there is active absorption and root pressure. Just as root pressure is not the most important method of water translocation in plants, active absorption does not appear to be the most important method of water absorption.

Passive Absorption. Passive absorption accompanies shoot tension and is a result of it. Not only is water pulled up the xylem, but it is apparently also pulled across the tissues of the root from the soil. Since the water in the xylem is under a tension or negative pressure, it has a lowered diffusion pressure and this could create a steep diffusion pressure gradient across the tissues of the root. However, the water may not only diffuse through the root but may also flow across it through what is called the **free space** as it is placed under tension. The free space of a tissue includes the cell walls and the cytoplasm, but not the vacuoles. Thus, water may be flowing through the walls and cytoplasm of one root cell to the next until it reaches the xylem, where it continues to flow on up, rather than moving through the vacuole of one cell to another.

Living root cells are not essential for passive absorption since water is just pulled through them by forces originating in the shoot. As a matter of fact, if the root system of a potted plant is killed by immersing the pot in boiling water briefly, the rate of passive absorption will increase considerably since the dead root cells now offer less resistance to the flow of water through them. However, this can go on for only a short period of time, for when the root cells are dead they soon begin to decay and disintegrate and water movement through them halts. Passive absorption can also go on without a root system at all if the cut end of the stem is immersed in water and if no air bubbles have entered the xylem. Recutting the stems (such as those of flowers for a bouquet) under water removes the portion of the stems into which air entered after cutting, facilitating passive absorption of water and making the flowers recover from wilting rapidly and last longer. Just as shoot tension appears to be the main method of water translocation, the passive absorption linked with it appears to be the main method of water absorption.

Soil Water. A consideration of water absorption by plants would be incomplete without some reference to the way water is held by soils and the relation of soil water to absorption. The water in soils consists of three main fractions held by different forces: 1. hygroscopic water; 2. capillary water; 3. gravitational water.

Hygroscopic water is a thin film held firmly on the surfaces of the small soil particles. It has a very low diffusion pressure and cannot be absorbed by plants. **Capillary water** is held by capillary action in the small spaces between soil particles, and its diffusion pressure is quite high, being reduced by only about one atmosphere by the presence of dissolved mineral salts in it. The capillary water almost always has a higher diffusion pressure than the water in plant roots and so can be absorbed by plants and provides the main source of water to plants. The capillary forces hold the water against the force of gravity. **Gravitational water** occupies the larger spaces between soil particles after a rain or irrigation, but within a matter of hours it drains out of the soil by the action of gravity or moves into the capillary spaces of the underlying dry soil. The spaces in which gravitational water occurs are too large for effective capillary action. Like capillary water, gravitational water is available to plants.

However, when gravitational water occupies the larger spaces in a soil it displaces the air from them and so the roots of plants have a deficient supply of oxygen. Many plants cannot thrive under these conditions of poor soil aeration and may even die. Tomato and corn are among the plants most affected, and in low and poorly drained parts of corn fields the plants are usually very small and have yellowed leaves. However, some plants such as rice, willow trees, and cattails thrive in waterlogged soil. For most plants the best soil water condition exists when the soil holds all the capillary water it can but no gravitational water, since there is then both adequate water and adequate oxygen available.

When a soil holds all the capillary water it can, but no gravitational water, it is said to be at **field capacity.** If the field capacity of a soil is 35% it means that the soil contains an amount of water equal to 35% of the dry weight of the soil when a capillary equilibrium is established. The percentage of water in a soil when all the capillary water is exhausted and only hygroscopic water remains is its **permanent wilting percentage,** for this is the point at which plants become permanently wilted. Thus, the field capacity sets the upper limit of capillary water in a soil and the permanent wilting percentage sets the lower limit. The difference between the two represents the percentage of capillary water the soil can hold and is called the **storage capacity** of the soil.

The field capacities (FC) and permanent wilting percentages (PWP) of different soils are quite unlike, both being influenced by the size of the soil particles. Clay soils have very small particles, silty soils larger particles, and sandy soils still larger ones. In a pint of clay soil there is much more particle surface than in a pint of sandy soil because of the more numerous soil particles. A clay soil therefore holds much more hygroscopic water than a sandy soil and has a higher PWP. Whereas the PWP of a clay soil might be 12%, in a sandy soil it is likely to be around 3%. A clay soil has more numerous and smaller spaces between particles than a sandy soil, and so can hold more capillary water. The FC of a clay soil might be about 37%, and that of a sandy soil about 15%. In these examples the clay soil would have a storage capacity of 25% (37% — 12%) and the sandy soil a storage capacity of 12%

(15% —3%). A silty soil would be intermediate in all respects. From a standpoint of available water and retention of water a clay soil is, then, superior to a sandy soil, though sandy soils are better aerated and easier to cultivate.

If a dry soil is watered it comes to field capacity, the depth of the moist layer depending on the amount of water supplied. Beneath this moist layer at FC will be dry soil sharply separated from the moist soil. If more water is not supplied to the soil the moist layer will only temporarily be above FC since this soil is already holding all the capillary water it can. The added water will be carried by gravity down to the dry soil and will wet an added layer to FC. Thus, as long as dry soil underlies the moist soil the amount of water supplied determines how deeply the soil will be wet, but not how wet the soil will be. Of course, if much water is added and the soil is poorly drained the soil may become waterlogged or there may even be flooding. The deeper layers of soils for a short distance above the water table may contain capillary water that has risen from the water table by capillary action. Because of its smaller capillary pores, a clay soil will be wet higher than a sandy soil by capillary rise from the water table. Plants with very deep root systems may secure water from this underground source and thrive while nearby plants with shallow roots may be undergoing permanent wilting. However, roots never grow through a layer of dry soil searching for moist soil. As a matter of fact, roots cannot grow through soil at the PWP. The deep roots grow down to the water table source only when the entire layer of soil is moist.

As water is absorbed from the soil around a root tip the water content of the soil is reduced and may even approach the PWP while soil a little farther away still contains considerable capillary water. However, this capillary water will not move toward the root any more than it moves into the dry soil below the layer at FC. As a root grows it enters a region of the soil that may not have been depleted of water (and minerals), and so continued root growth may be quite advantageous to the plant. However, it should be stressed again that the plant is not searching for water as this would be intelligent and purposeful behavior quite beyond the capacities of any plant.

Chapter 6

PLANT ACTIVITIES RELATED TO MINERAL ELEMENTS

From the soil (or from the water) in which plants are growing the plants absorb a variety of essential mineral elements as well as water. In addition to these substances green plants require only two other substances from their environment: the carbon dioxide used in photosynthesis and the oxygen used in respiration. That plants must have certain mineral elements if they are to grow well is known by everyone who has raised farm or garden plants, who has cultured plants by hydroponics (to be discussed later in this chapter), or who has even tended a few potted plants. Most people, however, think that when they are supplying fertilizer to plants they are providing plant foods, and the labels on fertilizer containers and advertisements for fertilizers help to further this misconception. Important as mineral elements are to plants, they are not really plant foods, as will be pointed out in the next chapter.

In ancient and medieval times plants were generally assumed to obtain their substance from the soil, as it was the only obvious source of the materials composing a plant. In the early seventeenth century, however, van Helmont, a Flemish scientist, performed a simple experiment that showed that plants do not obtain most of their materials from the soil. He planted a 5-pound willow tree in 200 pounds of soil and after five years found that the tree weighed 169 pounds while the soil had lost only two ounces. He concluded that plants are composed entirely of water, and although this was not a correct conclusion it was evident that plants do not get their food from the soil. He ascribed the slight loss in weight of the soil to experimental error. Actually, the mineral elements absorbed from the soil probably weighed somewhat more than the two ounces loss in weight.

THE ESSENTIAL ELEMENTS

Only a few of the 103 chemical elements are known to be essential for plant growth. They are generally the same ones essential for animals, though animals require a few that are apparently not essential for plants. The following memory device has been developed for remembering the essential elements:

C. HOPKINS CaFe–Mg CuZn–Na Cl–B Mn–Mo Co

This is read as "C. Hopkins cafe, mighty good cuisine, naughty clerk, burly manager, motley crowd." The letters are the chemical symbols of elements essential for either plants and animals. For plants, the "I" must be removed from Hopkins since iodine is not known to be essential for plants, and it is not certain whether sodium (Na) and cobalt (Co) are essential for higher plants or not. While chlorine (Cl) is essential for some species, it may not be generally required by all higher plants. After deleting these we have the following definitely established essential elements remaining:

C—carbon
H—hydrogen
O—oxygen

P—phosphorus
K—potassium (potash)
N—nitrogen

S—sulfur
Ca—calcium
Fe—iron
Mg—magnesium

Cu—copper
Zn—zinc
B—boron
Mn—manganese
Mo—molybdenum

Another value of the memory device is now obvious. The first three elements (C, H, and O) are those present in the largest quantities and are mostly derived from the carbon dioxide and water used in photosynthesis. They are not mineral elements derived from the soil. The next three (P, K, N) are the next most abundant in plants (though N should head the list in this respect) and are the elements absorbed in largest quantities from the soil. These three are present in what is called **complete fertilizer.** The next group of four elements (S, Ca, Fe, and Mg) are absorbed and used in still smaller quantities, though the amounts needed are still fairly substantial. Finally, the last five elements (Cu, Zn, B, Mn, and Mo) are used in only very minute quantities and so are known as **trace elements,** in contrast with the **major elements.** Any very large quantity of these trace elements in the soil will be toxic to plants, but small amounts are essential for proper growth.

A good fertile soil must contain adequate quantities of the seven major elements and the five trace elements. If plants are raised in hydroponic cultures all these elements should be present in the water supplied to the plants.

The elements are actually present in the soil as components of salt molecules, as many as seventy-five or more different salts usually being present. However, these salts are mostly highly ionized and it is both more convenient and more accurate to think of ions rather than the numerous salts they can form. Among the common ions present in the soil are phosphate (PO_4^{---}), potassium (K^+), nitrate (NO_3^-), ammonium (NH_4^+), sulfate (SO_4^{--}), calcium (Ca^{++}), iron (Fe^{++}), and magnesium (Mg^{++}). Any of these positively charged ions (**cations**) may combine with any of the negatively charged ions (**anions**) and form salts.

MINERAL NUTRIENT SOLUTIONS

Mixtures of the essential mineral elements, either as crystals or in solution, are available commercially for those who wish to engage in hydroponics, use foliar sprays, or use the so-called "concentrated" or "instant" fertilizers on their garden plants or potted plants. If they are used in quantity and one wishes to be certain that all essential trace elements are included, it may be cheaper and better to purchase the individual salts and make up a com-

plete nutrient solution at home. One formula follows:

1. Make up a 1-molar solution (the gram molecular weight of the salt dissolved in enough water to make a liter of solution, or perhaps half this amount of solution and salt) of each of the following salts: magnesium sulfate ($MgSO_4 \cdot 7H_2O$), calcium nitrate [$Ca(NO_3)_2 \cdot 4H_2O$], and potassium dihydrogen phosphate (KH_2PO_4). These are stock solutions.
2. Make up a liter of iron solution by adding 5 grams ferric chloride ($FeCl_3$) and 5 grams tartaric acid, or use 5 grams of an iron chelate.
3. Make up a trace element solution by adding the following to a liter of water: 2.9 grams boric acid (H_3BO_3), 1.8 grams manganese chloride ($MnCl_2 \cdot 4H_2O$), 0.1 gram zinc chloride ($ZnCl_2$), and 0.05 gram copper chloride ($CuCl_2 \cdot 2H_2O$).
4. The nutrient solution to be supplied to the plants is made up from the above stock solutions by adding the following quantities of each solution to about 500 c.c. of water and then adding enough more water to make a liter of solution: 2.3 c.c. of magnesium sulfate, 2.3 c.c. of potassium dihydrogen phosphate, 4.5 c.c. of calcium nitrate, 1 c.c. of iron solution, and 1 c.c. of trace element solution.

USES OF MINERAL ELEMENTS BY PLANTS

The principal use of mineral elements by plants is in synthesizing various compounds that are essential for the structure or the chemical processes of the plant. For example, nitrogen is an essential component of the molecules of amino acids, proteins, chlorophyll, alkaloids, purines, pyrimidines, and many other compounds. The major use of nitrogen is in making proteins. Phosphorus is a component of phospholipids, phosphorylated sugars, ATP, DPN, DNA, RNA (see Chapter 7), and many other very important compounds. Sulfur is a component of proteins and some other substances. Calcium is necessary for making the calcium pectate of the middle lamellae of cells. Iron is a constituent of some important enzymes. Each molecule of chlorophyll contains an atom of magnesium.

Mineral elements may also play catalytic roles other than as components of enzyme or coenzyme molecules. Some serve as enzyme activators, some possibly as coenzymes, and some may simply act as inorganic catalysts. It is likely that all the trace elements serve in some catalytic role, as indicated by the very small amounts of them that are required.

Ions as such have a variety of important influences on the activities of cells. For example, Ca^{++} ions decrease membrane permeability while K^+ ions increase it, a proper balance of the two being essential. Also, both the ions and molecules of salts influence the diffusion pressure of water in cells. These various roles of mineral elements are quite different from the roles of foods (carbohydrates, fats, and proteins) that will be discussed in the next chapter.

MINERAL DEFICIENCY SYMPTOMS

Plants that lack adequate quantities of one or more essential elements develop characteristic mineral deficiency symptoms. Since it requires much knowledge and practice to become proficient at diagnosing specific mineral deficiencies and to distinguish them from plant disease symptoms, we shall not go into detailed descriptions of the deficiency symptoms for each element. We shall, however, give the general types of deficiency symptoms.

Stunted growth is one of the most common symptoms, particularly if nitrogen, phosphorus, potassium, or calcium are deficient. The stems are also often unusually slender and woody.

Chlorosis, or a deficiency of chlorophyll and so a yellowish color, is a symptom of nitrogen, magnesium, iron, sulfur, and manganese deficiencies, each having its own characteristic patterns of chlorosis. **Necrosis,** or localized death of tissue, is particularly characteristic of phosphorus, calcium, potassium, iron, manganese, and boron deficiencies. The dead spots and terminal buds of boron-deficient plants turn a characteristic black color, while other deficiencies usually result in brownish necrosis.

Accumulation of anthocyanins in plants that do not usually produce it, and resulting in a red color, indicates a deficiency of nitrogen. Plants with a phosphorus deficiency have an unusual dark blue-green color, while plants with a potassium deficiency have a dull green color. Deficiencies of the major elements also usually interfere with fruit development, and may prevent it.

Another way of checking on mineral deficiencies in a soil is to test either the soil or plants growing in it (preferably both) for the amounts of each element present. Such chemical analysis can only be done by an expert, but many state agricultural experiment stations will test soil samples for citizens of the state at little or no cost. Kits for semi-quantitative color analyses for nitrogen, phosphorus, and potassium in soils or plants are available at some of the larger garden stores and it is not hard to learn to use them. Such chemical analyses may be more reliable than mineral deficiency symptoms.

THE ABSORPTION OF MINERAL SALTS

Land plants absorb most of their mineral salts from the soil through their roots, though they can absorb them through their leaves. Such foliar applications of mineral salts is becoming a common practice, and is particularly useful if an element tends to become tied up in an unavailable form in the soil. It cannot be used for calcium, however, since it can be translocated only upward in a plant.

The concentration of the various mineral ions in the cells of roots is generally considerably higher than in the soil, yet the ions continue to enter the roots. Obviously, something other than diffusion is involved, for the ions are moving from a region of higher to lower diffusion pressure. It has been found that such **accumulation** of mineral ions in the vacuoles of cells requires expenditure of energy from respiration, so the plant really does work in moving them in a direction opposite to that in which they would diffuse and also in holding them in the cells. When a cell is killed the ions promptly diffuse out of it.

It now seems likely that the minerals accumulated in the vacuoles of root cells remain there more or less permanently, and that most of the ions that go on up to the top of the plant pass through the free space of the root cells (the walls and cytoplasm), either by diffusing through or by being carried by the water flowing through the free space. Most mineral absorption occurs in the root tips and just back of them.

THE TRANSLOCATION OF MINERAL SALTS

Mineral salts are carried upward in plants through the xylem, being dissolved in the water that is flowing upward. In the younger parts of stems some minerals may also move up in the phloem. In general, minerals that enter the phloem are carried down. Thus, there may be something of a circulation of mineral salts in a plant. However, calcium cannot be carried downward in the phloem, and some elements like boron are not readily redistributed in a plant. Of course, mineral ions can diffuse from cell to cell and readily pass back and forth between the xylem and phloem.

Mineral salts may be washed out of the leaves of plants by rains, but in general mineral elements that have been absorbed remain in the plant until it dies, or loses leaves or other organs by abscission.

FERTILIZERS AND THEIR USE

As has been noted, fertilizers supply mineral elements rather than plant food, and the term fertilizer can be applied to any substance that contains substantial quantities of one or more of the essential elements. Fertilizers may be grouped roughly into three general classes: 1. organic fertilizers such as barnyard manure, green manure (crops plowed under), and compost; 2. commercial or "chemical fertilizers" sold in large bags in powder or pellet form; 3. concentrated or "instant" fertilizers that are relatively pure mixtures of mineral salts and must be dissolved in water for application. Any of these types may be quite satisfactory.

The commercial fertilizers are referred to as complete when they contain nitrogen, phosphorus, and potassium, though they would not really be complete unless they contained all essential elements. Only a few of the most expensive fertilizers are complete in this sense. The percentage content of commercial fertilizers is given on the bag. Thus a 5–10–5 fertilizer contains 5% nitrogen, 10% phosphorus, and 5% potassium (potash). It is obvious that most of the fertilizer is inert material that serves as a carrier, though small amounts of some of the other essential elements may be present as impurities.

The "organic gardeners" believe that only organic fertilizers, and particularly compost, should be used in gardens, regarding commercial fertilizers as injurious to the soils, plants, and people who eat the plants. Most scientists feel, however, that these beliefs have no basis in fact, commercial fertilizers being quite satisfactory for use. The elements provided by them are just the same as those provided by organic fertilizers, although usually somewhat more abundant. Burning of plants by fertilizers is

a result of applying too much and consequently plasmolyzing the root cells; this can easily be avoided. Of course, if the soil is deficient in organic matter and requires more for the improvement of its structure, addition of organic matter is necessary, but this is different from the increasing of soil fertility by adding deficient essential elements.

On the other hand, organic fertilizers, and particularly manures, have certain disadvantages. They may harbor parasitic bacteria fungi, bacteria, or viruses that cause plant diseases and they also frequently contain abundant weed seeds. It should be stressed that plants properly fertilized with commercial or concentrated fertilizers will grow just as lush and will be just as nutritious as those provided with organic fertilizers. The practices of organic gardening are generally quite sound, but the theories are quite unacceptable and are based mostly on misconceptions and a poor understanding of the nature of plants and soils.

HYDROPONICS

The cultivation of plants in solutions of mineral salts rather than in soil is referred to as hydroponics. Sometimes sand, gravel, or cinders are used as a medium in which the roots can grow and anchor the plants, and in other cases the plants are supported in some way or another so that their roots are suspended in the mineral solution. In the latter type of setup air may have to be bubbled through the solution to provide the roots with sufficient oxygen. There are available several non-technical books on hydroponics that give detailed instructions for hydroponic cultivation of plants.

Some people engage in hydroponics as a hobby, and hydroponics is used rather extensively in the cultivation of greenhouse plants. It is not used much commercially, since it requires more trouble and work than soil cultivation in most cases. However, it was used by the armed forces during World War II to raise vegetables for troops stationed on coral islands where there was little or no soil, and in Asia where the soil was badly infected with worms and other human parasites.

Hydroponics is a relatively recent cultural technique, but plant physiologists have used similar techniques for the last hundred years or so to determine what mineral elements are essential for plants. The plants are raised in either pure quartz sand that has been thoroughly washed to remove all soluble salts or in the solutions alone; they are provided with all known essential elements except the one being checked on. In such experiments it is essential that the salts be very pure and that there be no contamination from the containers or other sources. If a plant lacking a single element does not grow properly the evidence indicates that this element is essential. Such techniques are also used to demonstrate the deficiency symptoms produced by

lack of a particular element and to determine the best proportions of the various elements for optimum growth of various species of plants.

The fact that beautiful crops of plants can be produced by hydroponics emphasizes the fact that the soil provides only water, mineral elements, and anchorage for plants and that a soil rich in organic matter is not essential for good plant growth. However, the plants raised by hydroponics are not really superior to those that can be raised in a good, fertile soil.

Chapter 7

PLANT ACTIVITIES RELATED TO FOODS

We all know how essential foods are to man and other animals, and that without an adequate supply of food, malnutrition and even starvation will result. It is not so commonly recognized that plants, too, depend on an adequate food supply and will starve if deprived of food. Many people who do realize that plants require food are confused as to the nature of plant food, considering that it is quite different from animal food and is absorbed by plants from the soil. This misconception about food for plants no doubt comes about to a considerable extent from the practice of labeling fertilizers as "plant food." As has been pointed out in the previous chapter, van Helmont showed long ago that plants do not get their food from the soil, and despite the importance of mineral elements to plants, these elements are not really food. Of course, clearing up this confusion depends on just what biologists mean by the term "food."

WHAT ARE FOODS?

Though biologists are not in complete agreement as to how "food" should be defined, practically all botanists and the great majority of zoologists consider foods to be organic substances that can be used by living organisms either as a source of energy or in building the cells of which the organisms are composed. In other words, **foods are substances that can be used either in respiration or assimilation.** When foods are defined in this manner it becomes quite clear which substances qualify as foods: carbohydrates, fats, and proteins (along with somewhat simpler organic compounds such as glycerine, fatty acids, amino acids, and other organic acids from which one or more of these three classes of foods are made or into which they may be broken down). These substances are the foods of all living things, whether green plants, animals, or non-green plants such as fungi and bacteria.

Excluded as foods by this definition are water, mineral salts, and vitamins, even though all three are essential for all kinds of living organisms. While water is used in the construction of cells, it is not oxidized in respiration with a release of energy. The same may be said of mineral elements and vitamins. The many essential roles of mineral elements in plants were outlined in the previous chapter and are quite different from the roles played by foods. Vitamins play roles similar to those of hormones, the two classes of compounds being distinguished largely by the fact that animals make their own hormones but generally not vitamins. Since green plants make both their own hormones and vitamins, it is difficult to draw a line between them as far as plants are concerned.

A small minority of biologists prefer to define foods as essential substances obtained by living organisms from their environment. Aside from the fact that this definition makes oxygen a food (something neither laymen nor most biologists would really want to concede), the main objection to this definition is that various groups of organisms would have different kinds of foods. The foods of animals would include not only carbohydrates, fats, and proteins, but also water, vitamins, mineral salts, oxygen, and even carbon dioxide. The foods of green plants would consist only of water, mineral salts, oxygen, and carbon dioxide. These substances would also be foods for bacteria and fungi, but utter confusion would result as to whether carbohydrates, fats, proteins, and vitamins were foods for these groups of plants. Some kinds of bacteria and fungi can make all their own vitamins, other kinds can make some, and still others none. Some kinds can get along without an outside source of proteins, while others cannot. And then there are the nitrogen-fixing bacteria and algae. For them, but for no other organisms, nitrogen gas would be a food.

It is to avoid such confusion as to what is meant by "food" that we go along with the great majority of biologists in defining foods as organic substances that may be used in respiration or assimilation, regardless of whether they come from outside sources (as in animals and non-green plants) or are made internally (as in green plants). Whenever the term "food" is used in this book it is used in this sense.

PHOTOSYNTHESIS

Green plants (that is, plants containing chlorophyll) make their basic food by the process of photosynthesis. This process may be represented by the following summary equation:

$$\underset{6CO_2}{\text{carbon dioxide}} + \underset{6H_2O}{\text{water}} \xrightarrow{\text{light}} \underset{C_6H_{12}O_6}{\text{sugar}} + \underset{6 O_2}{\text{oxygen}}$$

Though this commonly used equation is not incorrect as a general overall summary, we shall see later that it does not tell much about how the process of photosynthesis really goes on, and it also gives a number of false impressions. For example, the carbon dioxide and water do not react with one another in photosynthesis. When they do react, the product is carbonic acid (H_2CO_3) and not sugar and oxygen. For the present, however, the equation will serve our purpose.

The Importance of Photosynthesis. That photosynthesis is one of the most important processes on earth cannot be questioned, since it is the only process that makes food from substances that are not foods. (There is one exception: the chemosynthetic processes of a few species of bacteria. These processes are similar to photosynthesis, but the energy used comes from the oxidation of inorganic compounds rather than from light.) Thus, photosynthesis is the ultimate source of food, not only for green plants, but for animals and non-green plants as well. If all photosynthesis on earth should suddenly stop, it would be but a matter of time until there was universal starvation and the extinction of life on earth. (Again excepting those few chemosynthetic bacteria and the few small animals they could supply with food.)

Besides, photosynthesis is the only source of oxygen in the atmosphere, another reason why this process is essential for survival. Before photosynthetic plants appeared on earth there was no oxygen in the air. Though yeasts and a few other fungi and bacteria can survive in the absence of atmospheric oxygen by carrying on a type of respiration that does not use oxygen (anaerobic respiration), the vast majority of plants and animals suffocate and die in the absence of oxygen. (As we shall see later, even green plants use oxygen in respiration.) Though photosynthesis in the only natural process adding oxygen to the air, oxygen is removed from the air by many processes such as respiration, combustion, rusting, and weathering of rocks.

Photosynthesis is also the only natural process that converts any appreciable amount of the energy from the sun into energy usable by living organisms. The light energy absorbed by chlorophyll during photosynthesis is converted into the chemical energy of the sugar produced (and then the chemical energy of the sugar is transferred to other organic substances made from it). When sugar or other foods are oxidized in respiration, the chemical energy of the foods becomes available for use in muscle contractions, nerve impulses, growth, and other work essential in the life of animals or plants. When we say that photosynthesis is essential as the source of food, we are also saying that it is essential as the source of much, though not all, of the energy used by living things. (The final source of this energy is, of course, the sun.) Some of the energy used in living things comes from the energy of moving molecules, while a number of life processes other than photosynthesis use light energy directly.

All fuels such as wood, coal, gas, and petroleum contain chemical energy that can be traced back to light energy from the sun converted into chemical energy by photosynthesis. Wood, of course, comes from trees, while coal was formed from the remains of ancient plants. Whether natural gas and petroleum were formed from the tissues of ancient plants or those of ancient animals that lived on plants, they are derivatives of sugar once produced by photosynthesis. When we burn fuels in our furnaces, cars, locomotives, or steam turbine boilers, we are releasing energy that can be traced back to photosynthesis. While atomic energy and perhaps the direct use of energy from the sun may some day be important sources of energy for use in our homes, industries, and transportation, at the present time about 90 per cent of our energy comes from fuels and so from photosynthesis. Most of the remaining energy used comes from water power. Much of the energy from fuels, and practically all from water power, is converted into electricity, which we then use in many ways.

Man is, then, dependent on plants not only for the food and oxygen that are essential for his survival, but also for the fuels that have made possible our modern industrial civilization. In addition, of course, man is dependent on plants and animals—and so in the final analysis on photosynthesis—for many items of commerce other than foods.

The Magnitude of Photosynthesis. Photosynthesis is not only an extremely important process, but it is also the largest scale production process on earth. It has been estimated that the plants of the earth produce about 380 billion tons of sugar each year by photosynthesis. This means that 555 billion tons of carbon dioxide and 230 billion tons of water are used, while 405 billion tons of oxygen are produced. The amount of light energy converted into the chemical energy of the sugar is about 1.28 million billion large calories. Despite the immense amount of energy from the sun trapped by photosynthesis, it is worth noting that only about 0.18 per cent of the sun's energy reaching the earth is used in photosynthesis.

About 90 per cent of the total photosynthesis on earth is carried on by marine algae living in the oceans, so the total sugar produced by land plants in a year is only about 38 billion tons. About a fourth of this is produced by crop plants. The estimate of total world photosynthesis given here is one of the most conservative, other estimates running up to as much as 580 billion tons of sugar per year.

The Nature of the Photosynthetic Process. Up to about the time of World War II very little beyond the summary reaction was known about the photosynthetic process, though it was realized that photosynthesis occurred in a series of steps and that the summary equation really gave little information as to just what was going on. When radioactive carbon (C^{14}) became available after the war for use as a tracer, rapid progress was made in working out the details of photosynthesis. Somewhat earlier, the use of a non-radioactive oxygen isotope (O^{18}) made it possible to determine that all the oxygen produced by photosynthesis comes from water, rather than from CO_2 as had previously been thought. The process of photosynthesis has now been worked out in detail as a result of intensive research during the

past quarter century. Though the details are still subject to some revision as a result of further research, we now have a reasonably satisfactory picture of the process. We now know, too, that the entire process of photosynthesis occurs within chloroplasts (Fig. 65), rather than involving in any direct way the other regions of a cell.

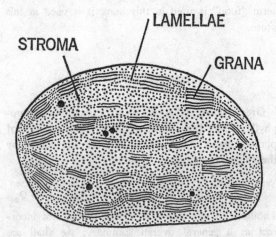

Fig. 65. Drawing of a chloroplast with the fine internal structure revealed by an electron microscope. Note the groups of dark grana, resembling stacks of coins, that contain all the chlorophyll, and the thin lamellae connecting the stacked grana. The remainder of the space is filled with stroma. The small black dots are probably lipoid bodies. The light reactions of photosynthesis occur in the grana and the subsequent steps in the stroma.

Photosynthesis begins with the absorption of light energy by chlorophyll. The chlorophyll is now "excited" or at a higher energy level; or, to be more specific, electrons in the chlorophyll have been moved to a higher energy level. The energized electrons are capable of doing work, and accomplish the following three things: 1. Water molecules (H_2O) are split into hydrogen and oxygen. The oxygen comes off as a gas, but the hydrogen is transferred to a hydrogen acceptor as outlined in the next step. 2. The hydrogen acceptor present in the chloroplasts is triphosphopyridine nucleotide (TPN), a rather complicated organic substance. When the TPN picks up the hydrogen from the water it is reduced chemically, and so is at a higher energy level. It is now $TPNH_2 \cdot$ 3. Another substance present in the chloroplasts is adenosine diphosphate (ADP). The electrons provide the energy necessary for adding a third phosphate group ($-H_2PO_3$, commonly represented as (P)) to the ADP, converting it to adenosine triphosphate (ATP). The ATP is a high energy compound that can readily give up its energy to other compounds and so do useful work. The reactions of these first steps in photosynthesis may be summarized as follows:

water	triphospho- pyridine nucleotide	phosphate	adenosine diphosphate
12 H_2O +	12 TPN +	18 P +	18 ADP

oxygen	reduced TPN	adenosine triphosphate
$\longrightarrow$ 6 O_2 +	12 $TPNH_2$ +	18 ATP

The net result of this first series of photosynthetic reactions has been to convert the light energy into chemical energy of the $TPNH_2$ and the ATP. (The symbols ATP and TPN as well as (P) are just abbreviations for the compounds and are not chemical symbols as, for example, C, H, and O are.)

The second main sequence of reactions in photosynthesis begins with the incorporation of carbon dioxide in a five-carbon phosphorylated sugar (ribulose 1,5-diphosphate). (Carbon dioxide fixation is not limited to green plants, but occurs in all types of plants and animals.) The resulting six-carbon compound is very unstable, and immediately breaks down into a three-carbon compound, phosphoglyceric acid. At this point the ATP from the first series of reactions enters into the reaction, contributing phosphate that converts the phosphoglyceric acid to diphosphoglyceric acid, a compound of higher energy content. In the next step the $TPNH_2$ contributes its hydrogen, converting the diphosphoglyceric acid into a three-carbon phosphorylated sugar (triose phosphate). The connections between the first and second series of photosynthetic reactions are now complete, and so the triose sugar may be regarded as the first product of photosynthesis. The original light energy has been passed on from the $TPNH_2$ and the ATP to the triose and is present in the triose molecules as chemical energy.

The triose molecules do not last long, however.

Of every twelve triose molecules, six are converted to three molecules of six-carbon phosphorylated sugar (fructose 1,6-diphosphate). One of these three molecules of fructose is converted to non-phosphorylated six-carbon sugars such as fructose or glucose, and in turn starch or sucrose may be made from these sugars. This one molecule of phosphorylated fructose (per twelve triose molecules produced or six carbon dioxide molecules used) represents the net product of photosynthesis. The other two molecules of fructose (2 C_6) plus the remaining six molecules of triose (6 C_3) are converted back to six molecules of the ribulose diphosphate (6 C_5), which is now ready to pick up some more CO_2 and start another cycle ($2C_6 + 6C_3 = 6C_5$, or 30C=30C).

This second series of reactions may be summarized as follows:

1.
carbon dioxide	ribulose 1,5- diphosphate	phosphoglyceric acid
6 CO_2 +	6 $(RH)_2$	$\longrightarrow$ 12 RCOOH

2.
phosphoglyceric acid	reduced TPN
12 RCOOH +	12 $TPNH_2$ +

adenosine triphosphate	hexose sugar
18 ATP $\longrightarrow$	$C_6H_{12}O_6$ +

ribulose 1,5 diphosphate	TPN
6 $(RH)_2$ +	12 TPN +

adenosine diphosphate	phosphate	water
18 ADP +	18 P +	6 H_2O

Again note that $(RH)_2$ and RCOOH are simply shorthand and not chemical formulas. It must be stressed that all the above equations are simply summaries. To go into each step of the reactions in detail would involve more complications than we wish to discuss here, though the detailed reactions have been worked out quite satisfactorily.

The equations we have given may seem to bear no resemblance to the general summary equation for photosynthesis as given near the beginning of this chapter. However, when each substance which appears on both the left- and right-hand sides of the equations below is canceled out, the result is the old standard equation:

$$12 H_2O + 12 TPN + 18 ADP \longrightarrow 12 TPNH_2 + 18 ATP + 6 O_2$$
$$6 CO_2 + 6 (RH)_2 \longrightarrow 12 RCOOH$$
$$12 RCOOH + 12 TPNH_2 + 18 ATP \longrightarrow 6(RH)_2 + 12 TPN + 18 ADP + 18 P +$$
$$6H_2O + C_6H_{12}O_6$$

$$6 CO_2 + 6 H_2O \longrightarrow 6 O_2 + 6 C_6H_{12}O_6$$

If you have never studied chemistry (or perhaps even if you have) this section on the photosynthetic process may not be too clear to you. However, even if you only realize that the process is much more complicated than is indicated by the general summary equation, it will be worth the time you have spent reading it. The really important thing to remember about the photosynthetic process is that it converts light energy into chemical energy, first of the energy transport compounds $TPNH_2$ and ATP and then of the sugar. Remember, too, that the hydrogen of the sugar produced comes from water while the carbon and oxygen come from the carbon dioxide.

The P/R Ratio. Green plants, like animals, carry on the process of respiration using sugar and oxygen and producing carbon dioxide and water. While photosynthesis occurs in a green plant only in the light and only in those cells containing chloroplasts, respiration occurs through the twenty-four hours of every day in every living cell, including cells without chloroplasts such as those of the roots. If the total amount of respiration in a plant over a period of several days or more just equaled the total amount of photosynthesis the plant would be using all the sugar and oxygen produced by photosynthesis in respiration. No surplus food would be available for use in assimilation or accumulation, and the plant could neither grow nor accumulate food. From a broad biological standpoint, there would be no net addition of oxygen to the air nor any net food surplus that could be used by animals or non-green plants if all plants of the earth consumed just as much food in respiration as they produced in photosynthesis.

Actually, plants in general produce from four to ten times as much sugar by photosynthesis as they use in respiration. The amount of sugar produced in photosynthesis divided by the amount used in respiration during the same time period gives the **photosynthetic/respiratory ratio,** or P/R ratio. Thus, the P/R ratio for a whole plant over a time period such as a growing season would be from 4 to 10 or so. However, the P/R ratio may be calculated on other bases. For example, we might want the P/R ratio of a single leaf over a period of two hours in the light. Such a P/R ratio would be quite high. If the P/R ratio of a plant should be less than 1 over an extended period of time the plant would be starving.

Photosynthetic production can be measured in two different ways. The **net** photosynthesis is the amount of sugar produced less the amount used in respiration by the chloroplast-containing cells. **Total** photosynthesis includes all the sugar made, without deducting that used simultaneously in respiration. Actually, when the rate of photosynthesis is being measured, either on the basis of the CO_2 consumed, the O_2 produced, or the sugar produced, it is net photosynthesis that is measured. To get total photosynthesis it is necessary to place the plant in the dark and measure the rate of respiration. This is then added to the net rate of photosynthesis to get the total rate.

Factors Affecting the Rate of Photosynthesis. The rate of photosynthesis may be influenced by the intensity and wave length of the light, temperature, the availability of carbon dioxide and water, or by any factor that limits the production of chlorophyll or the essential enzymes or energy carriers (ATP and $TPNH_2$), and by several other factors of lesser importance that we shall not discuss here. In any particular plant (or perhaps better, in any particular chloroplast) only one of these is the **limiting factor,** and only by increasing it can the rate of photosynthesis be increased. Thus, in a certain case, the light may be quite dim and so is the limiting factor. As light intensity is increased the rate of photosynthesis increases in proportion up to a point where further increase in light intensity brings about no further increase in photosynthesis. At this point light has ceased being the limiting factor and some other factor is now limiting. What the new limiting factor is can be determined only by experimentation. If the plant is now supplied with additional CO_2 and the rate of photosynthesis then continues to increase with light intensity we know it was CO_2 that had become limiting. If adding CO_2 has no such effect we know that some other factor had become limiting. The principle of limiting factors applies to other processes in plants and animals as well as to photosynthesis.

In nature, light is of course the limiting factor in photosynthesis at night and also early in the morning, late in the evening, on very cloudy days, and in such situations as a dense forest or deep in a lake or ocean. In the latter situations the light is not only of low intensity, but it is also deficient in the red and blue wave lengths that chlorophyll absorbs best as a result of being filtered through the leaves of the trees in a forest or by the water in a lake or ocean. Only plants that survive with a rather low rate of photosynthesis are able to live under the trees in a dense forest. Of course, no photosynthetic plants can live farther down in an ocean or deep lake than light can penetrate.

During ordinary daytime light conditions, carbon dioxide is most commonly the limiting factor in photosynthesis. If the carbon dioxide content of the air were higher than the 0.03 per cent usually present, photosynthetic production would be much greater than it is. The CO_2 content of the air in the immediate neighborhood of plants carrying on photosynthesis may be reduced to an even lower concentration than the 0.03 per cent. For example, it has been found that the air around and above the corn plants in a field may be reduced to as little as 0.01 per cent by photosynthesis. Unfortunately, there is no practical way of adding CO_2 to outside air, though CO_2 is sometimes added to the air in greenhouses. One advantage of fertilizers not commonly recognized is that they may increase the population of soil fungi, bacteria, and animals and so increase the CO_2 produced by their respiration. This extra CO_2 may increase the rate of photosynthesis in low-growing plants. An interesting possibility with corn is planting a field with several rows of low-growing variety alternating with several rows of a tall-growing variety. The uneven height of the plants causes air turbulence, bringing air with the usual 0.03 per cent CO_2 to the plants.

Temperature is usually the limiting factor only in quite cool weather, as with evergreens in winter. As temperature increases the rate of photosynthesis increases more slowly than the rate of respiration, and may actually decrease at higher temperatures. Besides, temperature is more commonly the limiting factor in respiration than in photosynthesis. In hot summer weather the P/R ratio becomes quite low and may even become less than 1. This is one reason hot weather is not good growing weather for many kinds of plants. Maine and Idaho are good potato-growing regions largely because of their cool summer weather. The result is a high P/R ratio in the potato plants and so a large amount of surplus sugar that contributes to the growth of large tubers and the accumulation of much starch in them.

When plants are wilted, or are approaching wilting, water may be limiting photosynthesis. This is not so much because there is too little water for use in photosynthesis as because of the reduced water content of the cells. Most processes in plants go on quite slowly when the water content of the protoplasm is deficient. Besides, when plants have a water deficiency their stomata may close, thus greatly interfering with the diffusion of CO_2 into the leaves. In this case, CO_2 is really the limiting factor even though the CO_2 deficiency is a result of a water deficiency.

In nature, photosynthesis is rarely limited by lack of sufficient chlorophyll, enzymes, or energy carriers in the chloroplasts. However, a plant lacking certain essential mineral elements may have its photosynthesis limited by a deficiency of these essential substances. Magnesium is a component of the chlorophyll molecule, as is nitrogen, and without these elements chlorophyll cannot be synthesized. Other elements such as iron are also essential for chlorophyll synthesis, even though they are not part of the chlorophyll molecule. Iron, manganese, and other elements are necessary for the production of the various enzymes essential for photosynthesis, and since enzymes are proteins, nitrogen and sulfur are essential for their production. Phosphorus is a most important constituent of both $TPNH_2$ and ATP, and it is obvious why photosynthesis is limited by a lack of this element. Despite the importance of mineral elements in the production of chlorophyll, enzymes, and energy carriers, most plants in nature or in cultivation have sufficient minerals so that they do not become limiting factors in photosynthesis. In general, it is a deficiency of light, CO_2, or water that is limiting.

RESPIRATION

The term **respiration** has been used in a great many different ways: *e.g.*, as a synonym for breathing, for the diffusion of oxygen into and carbon dioxide out of cells or tissues, and for the oxidation of food that occurs in all living organisms with the release of energy from the foods. We are using the term in this last sense, the really basic biological one. Plants do not breathe (nor do animals that lack lungs or gills), but all plants and animals carry on respiration in their living cells throughout their bodies.

Despite the great biological importance of photosynthesis, it is no more essential than respiration. Without the energy released from foods by respiration an organism could not continue to live, for life involves a continuous expenditure of energy.

Energy Sources for Plants. The energy from foods released by respiration is used by plants in a variety of ways: in accumulating mineral ions, in the synthesis of various compounds such as fats and proteins, in the reduction of nitrates absorbed from the soil into ammonium compounds, in cell division, cell enlargement, and other aspects of growth, in assimilation, in the movement of flagella, and in many other ways. Plants do not use energy from respiration in the two ways that most of it is utilized in animals: muscle contraction and the generation of nerve impulses, the rate of respiration in plants being lower than it is in animals. If plants used as much food in respiration as animals do they would not have enough food left over from photosynthesis after they used other food in building their tissues (assimilation) to support the animals and non-green plants on the earth.

The energy used by plants in their activities also comes from two sources other than respiration. One is **light energy direct** (as contrasted with light energy incorporated in foods during photosynthesis, or light energy indirect) used in various processes such as chlorophyll synthesis, anthocyanin synthesis, and phototropism as well as in photosynthesis. The other is **molecular energy,** used in diffusion and processes dependent on it such as translocation and transpiration. It may be noted that in plants the translocation of at least water and minerals uses molecular energy, while in animals the flow of blood occurs at the expense of energy from respiration used in contraction of the heart muscles.

The Process of Respiration. The summary reaction for the process of aerobic respiration (*i.e.*, the usual type of respiration requiring oxygen) is as follows:

$$\underset{\text{sugar}}{C_6H_{12}O_6} + \underset{\text{oxygen}}{6\ O_2} \longrightarrow \underset{\text{carbon dioxide}}{6\ CO_2} + \underset{\text{water}}{6\ H_2O} + \text{energy}$$

This equation, however, gives little more real information about the long series of individual reactions that make up respiration than does the summary reaction of photosynthesis for the photosynthetic processes. While we cannot consider the complete series of reactions making up respiration and all the enzymes involved in them in this elementary discussion, we shall break the respiratory processes down into four main subsummary steps.

The first of these series of reactions is referred to as **glycolysis** and results in the conversion of sugar to pyruvic acid. This involves the removal from the sugar of four hydrogen atoms, which become attached to a hydrogen acceptor (A), specifically diphosphopyridine nucleotide (DPN):

$$\underset{\text{sugar}}{C_6H_{12}O_6} + \underset{\text{hydrogen acceptor}}{2A} \longrightarrow \underset{\text{pyruvic acid}}{2\ CH_3COCOOH} + \underset{\substack{\text{reduced}\\\text{hydrogen}\\\text{acceptor}}}{2\ AH_2}$$

Before the sugar is converted into pyruvic acid it is first phosphorylated and then converted into a series of compounds including 3-carbon sugars (trioses). Glycolysis results in the transfer of about 4 per cent of the energy in the sugar to the reduced hydrogen acceptor, the rest being in the pyruvic acid.

The next main set of respiratory reactions, known as the **Krebs cycle** or the **citric acid cycle** (Fig. 66), begins with the removal of CO_2 from the pyruvic acid, converting it to a 2-carbon compound related to acetic acid. This then combines with a 4-carbon acid (oxalacetic acid), forming the 6-carbon citric acid, which then is converted to a series of eight other acids by reactions that involve removal of CO_2 and hydrogen, the latter being taken up by hydrogen acceptors. At two different steps CO_2 is removed, the net result being a 4-carbon acid (oxalacetic acid). This is then ready to react with another molecule of pyruvic acid from glycolysis and so to go through the cycle again. The end result of the citric acid cycle is the splitting up of pyruvic acid into CO_2 and H and the attachment of the H to hydrogen acceptor molecules. Thus, all the energy that was contained in the chemical bonds of the sugar is now in the chemical bonds of the reduced hydrogen acceptor (the AH_2). Each turn of the cycle also consumes three molecules of water as shown below:

$$\underset{\text{pyruvic acid}}{CH_3COCOOH} + \underset{\text{water}}{3\ H_2O} + \underset{\substack{\text{hydrogen}\\\text{acceptor}}}{5A} \longrightarrow \underset{\text{carbon dioxide}}{3\ CO_2} + \underset{\substack{\text{reduced}\\\text{hydrogen}\\\text{acceptor}}}{5\ AH_2}$$

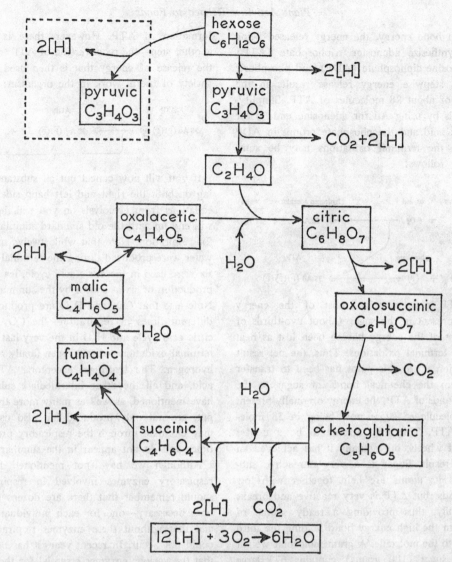

Fig. 66. Summary outline of aerobic respiration. Glycolysis converts sugar to two molecules of pyruvic acid, each going through the citric acid cycle like the one shown here. The net result of this is the breaking down of the pyruvic acid into 3 CO_2 and 10 [H]. The 10 [H] plus the 2 [H] from glycolysis are finally used in making 6 H_2O in the terminal oxidations, or a net of 3 H_2O subtracting the H_2O used in the citric acid cycle. Multiplying the 3 H_2O and the 3 CO_2 by 2 (for the other molecule of pyruvic acid from the sugar) gives the net product of respiration as shown in the general summary equation. Note that each step in the process balances. Several steps between citric acid and oxalosuccinic acid have been omitted for simplification.

Since two molecules of pyruvic acid are produced from one molecule of hexose sugar, the above equation should be multiplied by two. Thus, the product is 6 molecules of CO_2 (including all the carbon and oxygen originally present in the sugar) and 10 molecules of AH_2. Adding the 2 molecules of AH_2 produced in glycolysis, we have 12 molecules of AH_2 carrying all the hydrogen originally in the sugar (12 atoms) and the hydrogen from the water used (12 atoms). The AH_2 also contains all the energy originally present in the sugar.

The third series of reactions is known as **terminal oxidations**, and consists of the transfer of the hydrogen (and later just of electrons) from the AH_2 to a series of enzymes and finally to oxygen. At each transfer of hydrogen or electrons there is a decrease in bond energy, the energy released being used to synthesize adenosine triphosphate (ATP) from adenosine diphosphate (ADP) and phosphoric acid. This stepwise energy release results in the formation of about 38 molecules of ATP. Simplifying symbols by using Ad for adenosine and $\textcircled{P}$ for phosphoric acid and the phosphate groups in ADP and ATP, the terminal oxidations may be summarized as follows:

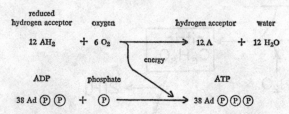

The ATP now contains most of the energy originally present in the sugar (about two-thirds of it), the rest of the energy having been lost as heat during the terminal oxidations. Thus, the net result of respiration up to this point has been to transfer energy from the chemical bonds of sugar to the chemical bonds of ATP, the energy originally present in one molecule of sugar now being in 38 molecules of ATP. This may appear to be a useless spinning of wheels, but actually it has achieved an important result. Sugar is a very unreactive substance and its atoms are held together with low energy bonds, but ATP is very reactive and breaks down readily, thus providing a ready source of energy from the high energy bond holding the third phosphate to the molecule. A gram molecular weight of hexose sugar (180 grams) contains 673 large calories and each high energy bond of ATP contains 12 large calories, or a total of 456 for the 38 gram molecular weights. The other 217 calories are converted to heat.

The same 673 calories would be released from 180 grams of sugar if it were burned, and the equation for the reaction would be the same as the summary equation for respiration. However, all the energy would be converted to heat and light (the flame) and the reaction would be quite violent and would require a high temperature. This is in such contrast with the gradual and controlled energy transfer from the sugar in respiration that respiration should never be confused with combustion, and we should not say that plants or animals burn sugar in their bodies.

Respiration is sometimes considered to end with the use of oxygen, the production of water, and the

formation of ATP. However, there is really still another step—the reconversion of ATP to ADP with the release of energy that is then used in doing a variety of useful work in the organism:

If you will now cancel out all substances appearing on both the right and left hand sides of all the equations from glycolysis on you will find that you will end up with the old standard summary equation for respiration. Note that while twelve molecules of water were produced during the terminal oxidations, six were used in the citric acid cycle, leaving the net production of six as shown by the summary reaction. Note also that CO_2 and H_2O are produced at quite different times in respiration—the CO_2 during the citric acid cycle and H_2O in the very last step of the terminal oxidations when oxygen finally reacted with hydrogen. The hydrogen acceptors, ATP, pyruvic acid, and all the other intermediate substances we have mentioned, as well as many more that we have not, are not only produced but also used, and so may circulate through the respiratory process many times and do not appear in the summary equation.

Although we have not mentioned the various respiratory enzymes involved in respiration, you should remember that there are dozens of different ones necessary—one for each individual reaction—and that without these enzymes respiration simply could not occur. In recent years it has been learned that the various enzymes essential for the citric acid cycle and the terminal oxidations are all located in the tiny bodies of cells known as **mitochondria,** probably in the same sequence as the reactions they activate. Thus, respiration is localized in specific cell structures, just as is photosynthesis.

Anaerobic Respiration. Most plants and animals carry on aerobic respiration such as has just been described, but yeasts and some other fungi, some bacteria, and even vascular plants when deprived of oxygen may carry on anaerobic respiration. This does not require oxygen in the air, and indeed some bacteria with anaerobic respiration cannot grow if oxygen is present. The most common type of anaerobic respiration is **alcoholic fermentation,** particularly characteristic of yeasts but also found for a time in higher plants lacking oxygen.

The first step in anaerobic respiration is glycolysis, just the same as in aerobic respiration:

$$\underset{\text{sugar}}{C_6H_{12}O_6} + \underset{\substack{\text{hydrogen}\\\text{acceptor}}}{2A} \longrightarrow \underset{\text{pyruvic acid}}{2\,CH_3COCOOH} + \underset{\substack{\text{reduced}\\\text{hydrogen acceptor}}}{2\,AH_2}$$

In the second step the pyruvic acid is converted to acetaldehyde and carbon dioxide:

$$\underset{\text{pyruvic acid}}{2\,CH_3COCOOH} \longrightarrow \underset{\text{acetaldehyde}}{2\,CH_3CHO} + \underset{\text{carbon dioxide}}{2\,CO_2}$$

In the third step the acetaldehyde reacts with the AH_2 and forms ethyl alcohol:

$$\underset{\text{acetaldehyde}}{2\,CH_3CHO} \quad \underset{\substack{\text{reduced}\\\text{hydrogen acceptor}}}{2\,AH_2} \longrightarrow \underset{\text{ethyl alcohol}}{2\,C_2H_5OH}$$

Summarizing the three steps we have the following general summary equation:

$$\underset{\text{sugar}}{C_6H_{12}O_6} \longrightarrow \underset{\text{ethyl alcohol}}{2\,C_2H_5OH} + \underset{\text{carbon dioxide}}{2\,CO_2}$$

In the course of these reactions enough energy is released to make two molecules of ATP from ADP and phosphate per molecule of sugar used. It is thus evident that anaerobic respiration is very ineffective in providing energy as compared with aerobic respiration: only 2 molecules of ATP instead of 38 molecules are produced per sugar molecule, or 24 large calories versus 456 calories per mol of sugar. The rest of the energy originally present in the sugar is now in the alcohol, and may be released by organisms that can oxidize it or by burning it as a fuel. The acetic acid bacteria regularly use alcohol in their particular type of aerobic respiration and release some of the energy from it:

$$\underset{\text{alcohol}}{C_2H_5OH} + \underset{\text{oxygen}}{O_2} \longrightarrow \underset{\text{acetic acid}}{CH_3COOH} + \underset{\text{water}}{H_2O}$$

This is what is happening when cider or wine "turns to vinegar." Since oxygen is required, growth of the acetic acid bacteria can be prevented by excluding air from the cider or wine.

When yeast is carrying on alcoholic fermentation and the alcohol content reaches about 14 per cent the yeast cells are killed by the alcohol they have produced. This is why the alcohol content of unfortified wines, non-distilled beverages like beer, and other unprocessed alcoholic beverages never exceeds about 14 per cent. Yeast carries on regular aerobic respiration when abundant oxygen is available.

Alcoholic fermentation occurs in higher plants principally in roots when the soil is flooded, depriving roots of oxygen. Although the roots of a few plants

can apparently survive for some time on anaerobic respiration, most roots will die after a few days or so. This may be either because of the accumulation of toxic quantities of alcohol or a deficient energy supply.

Factors Influencing the Rate of Respiration. Among the factors that affect the rate of respiration are temperature, oxygen, food supply, and water content of the tissues. Lack of water limits the rate in dry seeds or spores, the rate increasing greatly when water is supplied. In a wilted plant water may also be limiting. The food supply is practically never limiting except in a plant that is approaching starvation. Oxygen becomes limiting only when its concentration drops below the 20 per cent usually present in the air. In most plants most of the time temperature is the limiting factor, and as temperature increases so does the rate of respiration, up to the point (about 50° C.) where it becomes so hot that the respiratory enzymes are inactivated.

ASSIMILATION

It has already been noted that plants (and also animals) use foods in two basic ways: respiration and assimilation. It is impossible to discuss assimilation to the extent that respiration has just been discussed, partly because biologists still lack any very extensive or detailed knowledge of it and partly because what is known is mostly too technical for an introductory discussion.

Assimilation is the conversion of foods into essential cell structures, and so by extension the formation of the tissues and organs of an organism. Assimilation occurs not only during growth but also more or less continuously in living cells as they are being built up and torn down in something of a continuous renovation process. Sometimes assimilation is restricted to the formation of protoplasm, but since cell walls are essential parts of plant cells we are using the term to include also cell wall formation.

Assimilation may be considered to consist of two principal series of events: 1. the synthesis from the simpler foods of all the various substances such as proteins, nucleic acids, phospholipids, hormones, vitamins, enzymes, coenzymes, and pigments that are components of cell structures; 2. the assembly of these substances into the characteristic structural patterns of each cell. Much is known about the first series of events, though there is still much to be learned, and they are essentially biochemical in nature. It may be noted that cell walls are mostly cellulose, other carbohydrates, and carbohydrate-like substances. Protoplasm is mostly (around 90 per cent) water, the principal organic constituents being proteins, fats and other lipids, carbohydrates, and nucleic acids, though the less abundant constituents such as hormones and pigments are just as essential. Very little is known about the second series of events, though they are apparently basically a matter of organization rather than further chemical change. Just how the structure of the cells, tissues, and organs is controlled so that their assembly is characteristic of the species is not known in any detail, though it is certainly under hereditary control. This poorly understood aspect of assimilation involves problems in genetics, growth, development, and morphogenesis, areas of biology in which much research is in progress and which are slowly being clarified.

FOOD INTERCONVERSIONS

The sugar produced by photosynthesis is the source of all the other foods produced by plants and the chemical reactions involved have now been rather well worked out. In this introductory discussion we cannot go into these food syntheses in any great detail, but we shall outline a few of them. Plants not only synthesize more complex foods from the simpler ones, but they can also break the complex foods down into the simpler components.

Synthesis of Carbohydrates. The triose (3-carbon sugar) produced in photosynthesis is commonly converted into hexoses (6-carbon sugars) quite promptly, and plants can convert any of the many simple sugars (monosaccharides) present in them into any of the others through direct or indirect pathways. We cannot consider these reactions in any detail though most of them are known. Simple sugars are those that cannot be digested into smaller molecule sugars, though they can be converted to them in other ways. The simple sugars include not only trioses ($C_3H_6O_3$) and hexoses ($C_6H_{12}O_6$), but also sugars with other numbers of carbons such as pentoses ($C_5H_{10}O_5$) and heptoses ($C_7H_{14}O_7$). Each of these classes contains various specific sugars, differing from one another in the structure of their molecules rather than in the number of carbon, hydrogen, and oxygen atoms present. Among the important hexose sugars are glucose (dextrose), fructose (levulose), and galactose. The pentose sugars are not as abundant in plants as the hexoses, but are important in the construction of cell constituents. Ribose and deoxyribose (ribose lacking one oxygen or $C_6H_{12}O_5$) are essential constituents of the nucleic acids that are so important in cells, xylose and arabinose are used in making cell wall constituents, and we have already noted that ribulose is an important intermediate in photosynthesis.

All these sugars may be phosphorylated by the substitution of one or two phosphate groups into their molecules, and most reactions involving sugars can go on only when the sugars are phosphorylated.

Two molecules of simple sugars may unite, forming disaccharide sugars, and one of the simple sugars must be phosphorylated. Thus, sucrose is synthesized from a molecule of glucose-1-phosphate and a molecule of fructose:

$$\underset{C_6H_{11}O_6 \cdot H_2PO_3}{\text{glucose-1-phosphate}} + \underset{C_6H_{12}O_6}{\text{fructose}} \longrightarrow \underset{C_{12}H_{22}O_{11}}{\text{sucrose}} + \underset{H_3PO_4}{\text{phosphoric acid}}$$

If glucose were substituted for the fructose the product would be maltose, another disaccharide, or perhaps cellobiose (produced from a slightly different kind of glucose than is maltose). If galactose were substituted for the fructose the disaccharide produced would be lactose, or milk sugar. However, each of these reactions also requires a different enzyme (sucrose phosphorylase, maltose phosphorylase, etc.), and the synthesis can occur only if the organism can produce the proper enzyme. Thus, animals cannot synthesize sucrose because they lack sucrose phosphorylase, but they have lactose phosphorylase and so can make lactose.

Disaccharides may also be made from other monosaccharides including pentoses, but these are not common. Trisaccharides may also be made from three molecules of monosaccharides, tetrasaccharides from four and so on, but these sugars, too, are rather rare though some may be important. If many molecules of a monosaccharide are linked together the result is a **polysaccharide.** Like sugars, the polysaccharides are carbohydrates, but unlike sugars they are usually not soluble in water nor are they sweet. Among the common polysaccharides are starch, glycogen, and cellulose (all made from glucose), inulin made from fructose, and xylan and araban. The latter are cell wall materials made from xylose and arabinose, some of the pentoses. The number of monosaccharide molecules used in making a polysaccharide molecule ranges up to a thousand or more. The synthesis of a starch molecule consisting of 1000 glucose units can be summarized as follows:

$$\underset{\text{glucose-1-phosphate}}{1000\ C_6H_{11}O_6 \cdot H_2PO_3} \longrightarrow \underset{\text{starch}}{C_{6000}H_{10000}O_{5000}} + \underset{\text{phosphoric acid}}{1000\ H_3PO_4}$$

The enzyme activating this process is starch phosphorylase. The synthesis of other polysaccharides is similar, and indeed the above equation would also serve for cellulose synthesis, except that a different enzyme and a slightly different form of glucose are involved. Despite the great chemical similarity between starch and cellulose, however, they have quite different properties. Thus, starch collects in grains while cellulose forms strong fibers that are well-suited as cell wall components. Such differences go back to the fact that starch molecules are coiled and often branched, while in cellulose molecules the glucose units are arranged in straight chains. The cellulose molecules are bound together into fibrils, and these in turn into fibers.

Synthesis of Fats. Like the various carbohydrates, fats are synthesized from the triose sugar made in photosynthesis. The process occurs in three main sets of reactions. Some triose is made into glycerol (glycerine), an alcohol with the formula $C_3H_5(OH)_3$. It will be noted that this contains the same number of carbon and oxygen atoms as a triose molecule ($C_3H_6O_3$) but two more hydrogen atoms. The addition of hydrogen to a molecule is a reduction process that requires the expenditure of energy, and the energy for glycerol synthesis comes from respiration. Triose may also be converted into fatty acids, after essentially passing through glycolysis. To state the matter as simply as possible, an acetic acid (CH_3COOH) derivative is formed and then molecules of this are linked together forming larger fatty acid molecules: $CH_3CH_2CH_2CH_2CH_2CH_2CH_2COOH$. Each time two molecules are linked two oxygen atoms are removed. This, too, is reduction and requires energy, the source also being respiration. There are many fatty acids, depending on the length of the carbon chain, but because fatty acids are made from CH_3COOH there is always an even number of carbons. One common fatty acid is palmitic acid, $C_{15}H_{31}COOH$. It takes sixteen molecules of triose to make three molecules of palmitic acid and the number of carbon and hydrogen atoms come out even. However, the sixteen molecules of triose contain forty-eight oxygen atoms while the three fatty acid molecules contain a total of only six. Thus, there has been a removal of forty-two oxygen atoms, requiring a considerable expenditure of energy. This energy (and the energy used in making glycerol) is tied up in the chemical bonds of these molecules and both contain considerably more energy than carbohydrates.

The final step in fat synthesis is the reaction of glycerol and fatty acids:

$$\underset{\text{glycerol}}{C_3H_5(OH)_3} + \underset{\text{fatty acids}}{3\ C_{15}H_{31}COOH} \longrightarrow \underset{\text{fat}}{C_3H_5(C_{15}H_{31}COO)_3} + \underset{\text{water}}{3\ H_2O}$$

There are many different fats, depending on the kinds of fatty acids used, but all are made with glycerol. Generally the three molecules of fatty acid in a fat are different ones. The fats made by plants are usually liquid at room temperature and are called oils (such as corn oil or olive oil), in contrast with the solid fats from animals. The difference results from the fact that plant fats (oils) are to a large extent made from unsaturated fatty acids, that is, those that do not contain as much hydrogen as they could, e.g., $CH_3CH_2CH_2CH=CHCH_2CH_2COOH$. The unsaturated carbons form double bonds between themselves. If hydrogen is added to unsaturated fats, as it is in the manufacture of solid vegetable shortening or margarine, the fats become saturated and solidify.

Because of the high energy content of the glycerol and fatty acids from which they are made, fats also have a high energy content. Indeed, a pound of fat contains just about twice as many calories as a pound of carbohydrate, such as sugar or starch.

Synthesis of Proteins. Both carbohydrates and fats contain only carbon, hydrogen, and oxygen, but proteins contain in addition nitrogen and sulfur. These two elements must be added to the elements derived from photosynthesis when proteins are synthesized. The first step in protein synthesis is the synthesis of the twenty different kinds of amino acids from which proteins are built up. Some kinds of amino acids are made from others, but the basic sources of amino acids are some of the acids produced in the citric acid cycle of respiration. These are acids like pyruvic that contain a keto (CO) group. When an amino group (NH_2) becomes attached to the keto carbon an amino acid results. The simplest amino acid is glycine, $HCH \cdot NH_2 \cdot COOH$, but the other amino acids have the NH_2 and COOH groups that characterize the amino acid. In general their formula may be given as $RCH \cdot NH_2 \cdot COOH$, where R is a different grouping of atoms for each amino acid in place of the H of glycine.

The amino group is derived from ammonium compounds. Since plants generally absorb their nitrogen as nitrates such as potassium nitrate (KNO_3), conversion of the nitrates to ammonia involves both the removal of oxygen and the addition of hydrogen, a reduction requiring much energy from respiration. A few amino acids also contain sulfur.

Once the amino acids have been synthesized they are linked together into long coiled chains, thus forming proteins. Hundreds to thousands of amino acid molecules are required to make one protein molecule, so proteins as well as starch and cellulose have giant molecules. There are millions of different kinds of proteins, each species of plant and animal having its own particular kinds. Proteins may differ from one another in the kinds of amino acids they contain, the sequence of the amino acids in the chain, and the number of amino acid units in them, as well as in other respects. If even a single amino acid is out of place in a protein chain a new type of protein results, just as a change in a single letter of a word results in a different word or at least a misspelled word.

The problem of protein synthesis is more complex than that of starch or cellulose synthesis, where long strings of a single substance (glucose) are linked together. Since proteins are composed of many different amino acids arranged in a specific sequence, biologists have long been puzzled as to just how a plant or animal could put the amino acids together in just the right sequence to produce the particular proteins characteristic of the species. The situation is now beginning to be clarified. The code for protein synthesis seems to be provided by the hereditary potentialities or genes in the nucleus of the cell. The genes are apparently deoxyribonucleic acid (DNA), and this DNA serves as a matrix against which molecules of ribonucleic acid (RNA) are formed, with a copy of the DNA code. The RNA may then go from the nucleus into the cytoplasm as tiny bodies known as **microsomes.** It is in these microsomes that proteins are apparently synthesized, the RNA code determining the sequence in which the amino acids are assembled. Thus, hereditary control is exerted over protein synthesis, and the proteins characteristic of the species are formed.

Since enzymes are proteins and since enzymes determine what processes can occur in an organism, this precise control of protein synthesis is extremely fundamental and important since it determines what enzymes (as well as what other proteins) are formed. The genetic aspects of this hereditary control of protein synthesis will be discussed in Chapter 13.

Digestion. Plants not only build up complex foods such as proteins, fats, and starch, but they also break them down into the simpler foods from which they were formed—amino acids, fatty acids, glycerol, and simple sugars. These decompositions may be just the reverse of the synthetic processes and activated by the same enzymes, or they may involve different enzymes that promote the breakdown of foods by reaction of starch with phosphate, the product being glucose-1-phosphate, while amalayse catalyzes the reaction of starch with water, the product being glucose. The latter type of reaction is referred to chemically as **hydrolysis** (reaction with water) and biologically as **digestion.**

All digestion reactions are similar to the one just described. Thus, sucrose reacting with water under the influence of the enzyme sucrase breaks down into glucose and fructose, while maltose and cellobiose decompose into glucose. Fat, reacting with water, is digested into fatty acids and glycerol, while proteins are digested to amino acids.

In green plants digestion occurs within the various living cells, rather than in a digestive tract, as in most animals, or outside the body as in bacteria and fungi. The cell membranes are impermeable to proteins, fats, starches, and other complex foods, and so only after they have been digested can foods pass from one cell to another. Foods are also translocated through the phloem of plants in their simpler forms. Similarly, bread mold cannot absorb the starch, proteins, and fats of the bread on which it is growing, but after they are digested by enzymes secreted by the mold they diffuse into its filaments. The starch or fats accumulated in a seed cannot be used by the embryo plant until after they have been digested.

Practically all organisms have enzymes that digest proteins, fats, starch, and disaccharides, but most organisms lack cellulase, the enzyme that activated the digestion of cellulose, and so cannot use cellulose as food. Exceptions are the wood-decaying bacteria and fungi and the protozoa that live symbiotically in the digestive tracts of termites.

Digestive enzymes may be extracted from organisms without losing their activity. Several digestive enzyme extracts from plants are available commercially. Diastase, a crude extract from germinating grains or molds, contains amylase and maltase and so digests starch. It is available in most drug stores. The commercial meat tenderizers contain papain, a protein-digesting enzyme extracted from the fruit of the papaya tree.

FOOD ACCUMULATION

Any food left over from respiration and assimilation accumulates in the plant. The greater the excess

of food production by photosynthesis over the combined usage of food in respiration and assimilation, the larger the quantity of food accumulated. Animals accumulate food (and so get fat) when they eat more food than they use in respiration and assimilation. Practically all the food accumulated by animals is fat, but plants accumulate starch, proteins, and even sucrose (as in the sugar beet and sugar cane) as well as fat (Fig. 67).

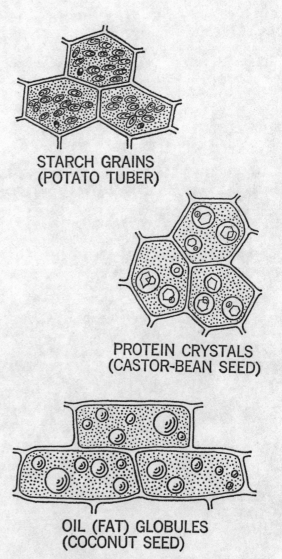

STARCH GRAINS
(POTATO TUBER)

PROTEIN CRYSTALS
(CASTOR-BEAN SEED)

OIL (FAT) GLOBULES
(COCONUT SEED)

Fig. 67. Foods accumulate in cells as grains (starch), crystals (proteins), and globules (fats).

Food may accumulate in almost any plant organ, but accumulation is most abundant in the endosperm and cotyledons of seeds, in fruits, in fleshy roots and tubers, in rhizomes, and in bulbs. The twigs of trees and shrubs contain considerable accumulated food. The large quantity of food accumulated in many seeds, such as those of legumes and grasses, makes them valuable as a source of food for humans. Plants do not store food for future use, but accumulated food is frequently digested and used, as when a seed germinates or when a twig resumes growth in the spring.

In addition to accumulating foods, some plants accumulate considerable quantities of other substances that they cannot digest and use. Among these substances are rubber latex, turpentine and resin, alkaloids such as atropin, digitalis, and nicotine, and essential oils such as peppermint and wintergreen. Although such substances are of no known value to the plants that make and accumulate them (as far as is known now) they may be of considerable commercial value.

Chapter 8

PLANT GROWTH

Everyone is aware of the fact that plants, as well as animals, grow and develop, but growth is pretty much taken for granted. When one stops to think about it, however, growth is really one of the most amazing and interesting things on earth. Biologists are able to describe the growth and development of a plant or animal in some detail, and know much about the influence of various environmental factors on growth. Nevertheless, just how a single-celled fertilized egg of a plant or animal grows and develops into a mature individual with the characteristic structural organization and behavior patterns of its species is still largely a mystery to biologists. Much research is going on in this area and advances in knowledge are being made, but there is still only a vague understanding of the controlling factors that cause certain patterns of cell division, cell enlargement, and cell differentiation to occur at certain times and certain places and so result in a particular kind of organism. The fertilized egg cell of a cabbage plant is very similar in appearance to that of a maple tree, but they develop into quite different kinds of plants.

It is obvious, of course, that heredity exerts a controlling influence on growth and development, and in Chapter 13 we shall consider what is known about the nature of this hereditary control. However, growth and development are also greatly influenced by the environment, as we shall see in the next chapter. The interactions between heredity and environment determine the course of the internal life processes of the organism, and these in turn determine the pattern of growth, development, and behavior. Although this general picture seems to be quite clear, little is known concerning the specific internal life processes involved in growth and development.

Development is sometimes considered as an aspect of growth and sometimes as distinct from growth. When the term "growth" is used in this more restricted sense it includes only the increases in size and weight, while "development" refers to the more qualitative cell specializations and the differentiation of tissues and organs characteristic of the species. However, growth and development are intimately intertwined, and growth without development occurs principally in some tissue cultures where masses of unspecialized cells are formed. Growth is generally measured by an increase in size or weight, while development is ordinarily described or illustrated. The weight may be either **fresh weight** (the weight of the living organism) or **dry weight,** the weight of the substances other than water. Of course, the dry weight can be determined only by killing the plant and drying it out. Thus dry weight can be determined only once for a particular plant, while fresh weight can be determined at intervals.

As an example of growth we may take a seed plant such as the common bean plant. The growth of a plant begins as soon as the egg has been fertilized by a sperm and this early growth results in the embryo plant within the seed. Growth then ceases until the seed is planted. The resumption of growth of the embryo is referred to as **germination.** The roots grow downward in the soil and soon the hypocotyl is enlonged, raising the fleshy cotyledons of the bean plant above the ground (Fig. 68).

Between the cotyledons is the plumule, or young stem tip with the first true leaves, and this proceeds to grow into the main stem of the plant, and later on to branch. The first pair of bean leaves is simple and opposite, while the remaining leaves are alternate and compound. After the stems and roots grow for several weeks, flower buds appear, flowers develop and are pollinated, and fruits (the pods) containing seeds are formed.

Bean plants, like most other annual plants, die after their production of mature seeds is completed.

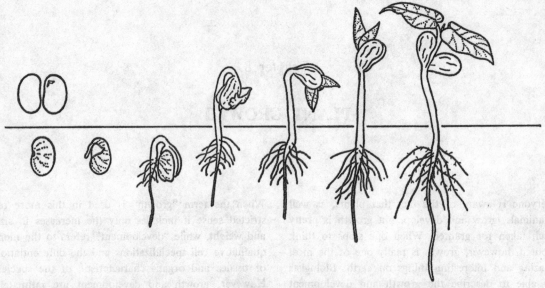

Fig. 68. Some stages in the growth and development of a lima-bean plant. The plant begins its life as a microscopic single-celled fertilized egg, which is not shown, but which develops into the embryo plant of the seed.

The growth of the plant is slow at first, then enters a period when it is very rapid, and then slows down again and finally ceases as maturity is reached. If the growth is plotted on a graph the growth curve is seen to have something of an S-shape. Such S-shaped growth curves are typical, not only for beans and other plants, but also for animals and even for populations of organisms.

THE CELLULAR ASPECTS OF GROWTH

Even though we measure the growth of an entire plant or of its various tissues or organs, we are in the final analysis dealing with what happens to the cells of the plant as growth occurs. At the cellular level growth involves three more-or-less distinct things: cell division, cell enlargement, and cell differentiation. Cell division results in an increase in the number of cells, but without cell enlargement there would be only a larger and larger number of smaller and smaller cells, so for an increase in size and weight both must occur. Furthermore, normal growth and development also require cell differentiation into the many kinds of cells making up a plant. Without such cell differentiation growth would result only in an increasing mass of parenchyma cells (as some-

times occurs in tissue cultures) and not in a recognizable plant with all its characteristic tissues and organs.

Cell Division. In vascular plants cell division occurs principally in the meristematic cells at the tips of the stems and roots and in the cambium, though cell division also occurs in young leaves, flowers, and fruits and to a greater extent than was formerly supposed in young stems back of the growing tip. The plane in which most of the cell divisions occur determines the general shape of the organ. Thus, in stem and root tips the new cells are produced mostly in a longitudinal direction, resulting in great elongation without a very great increase in diameter, while in the cambium most divisions occur in a radial direction and so result in an increase in stem or root diameter. In young leaves most cell divisions are in one plane, resulting in a flat structure. The characteristic shape of a leaf, whether entire or lobed, simple or compound, heart-shaped or ovate, is a result of a particular pattern of cell divisions (and to a lesser degree of cell enlargement) that is characteristic of the species. The same applies to flowers, fruits, and other organs. If cell divisions occur with about the same frequency in all planes spherical structures such as fruits result.

Cell division consists of two recognizable steps:

the division of the nucleus (almost always by a rather complicated process known as **mitosis),** and the subsequent division of the remainder of the cell (known as **cytokinesis).** When a nucleus divides by mitosis the first indication is a change in the appearance of the chromosomes, which become shorter, thicker, and more distinct (Fig. 69). The shortening and thickening of the chromosomes is at least partially a result of their coiling. Each chromosome duplicates itself before becoming shorter and thicker and can be seen to consist of the two strands (the chromatids) that result from this duplication, the strands frequently being intertwined. After this the nucleolus and nuclear membrane become indistinct and finally disappear and **spindle fibers** begin appearing in the cytoplasm at either end of the nucleus. Following the disappearance of the nuclear membrane they connect with each other through the region of the nucleus, forming the **spindle,** which resembles magnetic lines of force in appearance. These early events in mitosis, which require most of the time in a cell division, are referred to as the **prophase,** in contrast with the later stages of mitosis and with the cell when it is not dividing (the **interphase).**

In the next stage (the **metaphase)** the chromosomes line up along the equator of the cell. Soon the two chromatids of each chromosome separate from one another, one moving toward each end of the cell. This is the **anaphase.** Each chromatid is now a chromosome, the result being that each end of the cell has a complete set of chromosomes of the same number and the same kinds as were present at the beginning. Next the chromosomes begin to become longer and thinner, new nucleoli and nuclear membranes appear, and so two nuclei are constructed, each one with the same number and kind of chromosomes as the parent nucleus. This reconstruction of the nuclei is known as the **telophase,** and completes mitosis. In the meantime some of the spindle fibers appear to be contracting toward the equator and forming a cell plate. This splits and a middle lamella is secreted in the space between the two protoplasts. Following this, each of the new cells begins secreting a new cellulose wall against the common middle lamella and when these cellulose walls are laid down cell division is completed. There are now two cells in place of one.

The two cells are, however, no larger than the parent cell. Before another cell division the two cells generally enlarge until each one is the size of the parent cell. This cell enlargement is considered

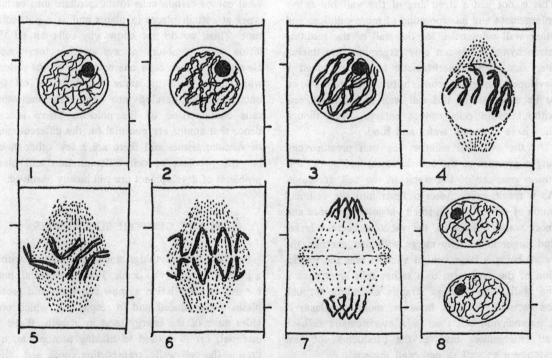

Fig. 69. Selected stages of mitosis and cytokinesis. 1–Interphase. 2–4–Prophase. 5–Metaphase. 6–Anaphase. 7–Telophase (early). 8–Daughter cells in interphase. See text for description.

an integral part of the cell division and should be contrasted with the much more extensive enlargement which occurs as the older meristematic cells become converted into parenchyma cells.

Mitosis may seem to be an unnecessarily complex way for nuclei to divide, but it achieves an important result: each new cell contains exactly the same number and kinds of chromosomes as the parent cell did. The reason that this is so important is that the chromosomes contain, in linear order, the genes or hereditary potentialities of the cell. Thus, each new cell produced by mitosis contains complete sets of genes. If a nucleus simply pinched in two by **amitosis,** as appears to occur in some lower plants and in a few tissues such as the endosperm of coconut seeds, the chances are that the resulting cells would not all contain complete sets of genes but might have extras of some and none of others.

Cell Enlargement. When meristematic cells become older they begin enlarging greatly, much more lengthwise than sidewise, until they may be five to ten times as long as they were or even longer. The result is a parenchyma cell, or a cell of a general parenchyma type that will differentiate into some specialized cell type. The primary step in cell enlargement is an increase in the size of the cell wall. This is not just a stretching of the wall but active enlargement and incorporation of more cellulose and other wall substances, for the wall of the resulting parenchyma cell is not only larger but also thicker than that of the meristematic cell from which it developed. The plant growth hormone **auxin** is known to be essential for this cell wall enlargement, and without auxin cells cannot enlarge, even though they have abundant water and food.

As the cell wall enlarges the wall pressure and turgor pressure of the cell decrease, and so the diffusion pressure of the water in the cell decreases. As a result, more water diffuses into the cell and much of this collects in small vacuoles. As more and more water diffuses in, the vacuoles become larger and larger and finally merge with one another, the result being a large central vacuole and the restriction of the cytoplasm to a rather thin layer against the walls, and perhaps strands extending through the vacuole. There is, however, more cytoplasm in a parenchyma cell than in a meristematic cell, so cell enlargement involves the production of new cytoplasm as well as new wall material.

After a cell has produced secondary cell walls no further enlargement is possible, since the cellulose fibers in secondary walls are arranged in rather rigid parallel bands rather than in a loose and random arrangement of fibers as in the primary wall.

Cell Differentiation. The production of cells such as sieve cells, epidermal cells, cork cells, fibers, tracheids, vessel elements, and guard cells involves the specialization of what are in the beginning essentially parenchyma cells. One of the first steps is frequently a change in size or shape, and in many cases secondary thickening of the walls is an essential step. This may occur only in the corners, as in collenchyma cells, in certain areas, as in the formation of vessel elements with ring or spiral thickenings, or all over (except at pits) as in fibers or stone cells. In the formation of cork cells the walls become impregnated with waterproof suberin, while epidermal cells secrete a layer of cutin on their outer surface. Cell differentiation may also involve the loss of certain structures, such as the end walls of vessel elements and the nuclei of sieve cells. The final step in the differentiation of cells such as tracheids, fibers, vessel elements, and cork cells is the death of the cells, and the disintegration of the materials that made up the protoplast, leaving only the cell walls.

While we can describe the events of cell differentiation in some detail, we do not know just what causes certain cells to differentiate into certain types at certain places in plants and at a particular time. Thus, we do not know why cells cut off the inside of a cambium always develop into vessel elements or other cells characteristic of the xylem, while those from the same cambium cut off the outside always develop into sieve cells and other cells characteristic of the phloem. There is evidence that auxins are essential for the differentiation of vascular tissue, and there are a few other clues, but the problem of cell differentiation and other problems of development are still mostly unsolved.

PLANT GROWTH SUBSTANCES

Several different things must be available within a plant if growth is to occur. Food is essential, both for use in assimilation as new cell walls and protoplasm are produced and in respiration, which provides most of the energy used in growth. Water is essential, for it is used in making protoplasm, hydrating the cell walls, transporting foods and other substances, and in maintaining an adequate degree of turgor pressure in the cells. Enzymes of many kinds are essential for activating the various biochemical processes involved in growth. Certain mineral elements are essential, primarily in the construction of various compounds. However, even if all these substances are present in plants in adequate quantities, growth cannot occur if another class of substances—the **plant growth substances** or **plant hormones** are not present in suitable quantities. These quantities are very small and far less than the requirements of water, foods, or even most of the mineral elements. A proper balance of the various plant growth substances is essential—too high a concentration as well as too low a concentration inhibiting growth in many cases.

The ratio between the concentration of two or more growth substances may be more important in some cases than their absolute concentrations. Unlike animal hormones, plant hormones are not produced in special ductless glands, but like animal hormones, they are produced in one part of the organism and transported to other parts, where they exert their influences.

The first class of plant hormones discovered were the **auxins,** and more research has been done on them than any others. They also appear to be the most basic of plant hormones and the most diverse in their effects. Auxin was not definitely identified as a hormone until 1928, but research leading to its discovery was begun by Charles Darwin in the last century. Another plant hormone is **traumatin,** or wound hormone, which is produced by injured cells and diffuses into nearby intact cells, causing them to resume meristematic activity and to produce a scar tissue that heals the wound. The **kinins** (particularly kinetin) appear to be hormones that are essential for cell division. The **gibberellins** are plant hormones that greatly promote stem growth and have other effects, but apparently can act only when auxins are also present. Most of the B vitamins also apparently act as plant hormones, since they are generally produced in one part of a plant and transported to other parts. Thus, thiamin or vitamin B_1 is produced in leaves but is essential in root growth. In animals vitamins can be distinguished from hormones, since the animals make their own hormones but must secure their vitamins along with their food.

Since plants make both their own vitamins and hormones this distinction cannot be made for plants. Indeed, neither vitamins nor hormones are a chemical class of compounds (as are proteins, carbohydrates, or fats, for example), but are diverse compounds grouped together for convenience because all of them have marked influences on organisms at low concentrations. It seems likely that both vitamins and hormones function primarily as coenzymes or are components of coenzymes. Animals and plants both apparently require the same B vitamins, but animal hormones are quite ineffective in plants and vice versa.

All the above plant growth substances have been identified, extracted from plants, and in many cases synthesized in laboratories. They are definite chemical compounds whose structures are known. In contrast with them is **florigen,** the hypothetical flowering hormone, which has never been extracted from a plant or identified chemically. There is, however, good evidence that such a hormone is produced in the leaves of plants and transported to the buds. There are no doubt many other plant hormones than the ones mentioned which still remain to be discovered.

A detailed consideration of all the various plant hormones is quite technical, but we shall consider two of them—the auxins and the gibberellins—in some detail here.

Auxins. The principal auxin of plants is a compound known as indole-3-acetic acid, but plants produce other compounds that have similar effects and so we should speak of auxins in the plural. Auxins are produced primarily in leaves (particularly the younger ones), pollen, and the developing embryo plants in ovules. Auxins can be transported only from the point of production downward, *i.e.,* from the tip of a stem toward the base. This is not a matter of gravity, for auxin travels only from the tip to the base even when a stem is inverted, as in the drooping branches of the weeping willow.

The basic influence of the auxins appears to be in the enlargement of cell walls, and in some unknown way the auxins are essential for channeling energy from respiration in this direction. Auxins also appear to be essential for cell division, though their role here may still be in the enlargements between divisions or perhaps in the formation of the new cell walls. The meristematic cells of the cambium in tree trunks do not begin dividing in the spring until auxin from the growing terminal buds reaches them, and other examples of the necessity of auxins for cell division are known.

Although auxins may have only one or two basic

roles, they have a multitude of varied effects on the observable growth, behavior, and development of plants. They are essential for the elongation of both stems and roots, although roots are much more sensitive to auxins than stems and a lower optimum auxin level. Too much auxin inhabits the growth of either roots or stems, but a concentration optimum for stem growth inhabits root growth. The growth of lateral buds is inhibited by auxins, and the auxins reaching them from the terminal bud are often concentrated enough to keep them from growing into branches. This is known as **apical dominance.** Some of the older buds farther away from the terminal bud eventually grow into branches when the auxins reaching them have been reduced to a low enough level, but some buds of trees never grow into branches and may become buried in the wood as a tree grows in diameter. Such buried buds are responsible for the grain of bird's-eye maple. One reason for pruning shrubs is to remove the apical dominance so that the shrub will be more highly branched and bushier. If a terminal bud is cut off and a dab of lanolin containing auxin is placed on the cut stem end the auxin prevents growth of the lateral buds just as the auxin from the bud did.

The auxins produced by leaves prevent the **abscission** (or falling off) of the leaves, abscission occurring when the balance of auxins reaching the petiole base from the leaf and stem sides is disturbed. If the blade of a leaf is cut off the petiole will soon abscise, but if auxin in lanolin is placed on the petiole stump abscission will not occur. The fall of leaves in the autumn results from a reduction in auxin production by the leaves, brought about by these factors. Premature fruit drop also results from deficient auxin production.

The development of ovularies into fruits is dependent on the production of auxins by the embryo plants in the ovules, and perhaps to a lesser degree on auxins from the pollen that pollinated the flower. In some plants such as seedless grapes and oranges the fruits develop even though there are no developing embryos or seeds (**parthenocarpy**), sufficient auxin apparently being provided by other sources. Unpollinated flowers can be made to develop into fruits by spraying them with auxin solutions (artificial parthenocarpy), such fruits being seedless.

Auxins are necessary for the formation of adventitious roots on stem cuttings. Some plants such as coleus and willow readily form roots on cuttings, but most species root well only when the stems are treated by dipping them in powder containing auxin or placing them in auxin solutions for a while. Cuttings not treated with auxin will sometimes root when the younger leaves are left on, but not when they are removed, the extra auxin provided by the young leaves being essential. Nurserymen propagate most of the plants they sell by cuttings and make extensive use of auxins in promoting root formation.

If a stem has a higher auxin concentration on one side than the other it will grow faster on that side, causing it to bend in the opposite direction. This result can be produced by applying auxin in lanolin to one side of a young stem. Both **phototropic** and **geotropic** bending result from unequal auxin concentrations. The more brightly lighted side of a stem has a lower auxin content than the less brightly lighted side, so it bends toward the light. If a plant is placed in a horizontal position auxin moves to the lower side of both the stem and the roots. This results in a more rapid growth of the lower side of the stem, so it bends and grows upward. The higher auxin content of the lower side of a root inhibits growth, and so roots bend and grow downward. These different geotropic responses of roots and stems are very important, as they insure that roots will grow down into the soil and stems upward into the air. It is consequently not necessary to plant a seed in any particular position.

Gibberellins. The effects of gibberellins on plants are perhaps more spectacular than those of auxins, but they are probably not as essential for growth as the auxins. The gibberellin content of different species or even of different varieties of a species may vary greatly, and many dwarf varieties differ from tall varieties of the same species in that they produce less gibberellin, or perhaps no gibberellin at all. By spraying a dwarf variety with a gibberellin solution it can be made to grow just as tall as the tall varieties. It appears that vines such as morning-glories and kudzus probably contain much gibberellin, while compact and slow-growing plants probably produce very little, if any.

In addition to their effect on stem elongation, gibberellins have several other influences on plant growth. Plants which normally bloom only when the daylight period is long (Chapter 9) will bloom in short days when treated with gibberellin. Biennial plants do not produce tall stems and bloom in nature the first year of their lives because they require a period of low temperature before such development can occur. However, such biennials will produce tall stems and bloom without cold treatment if supplied with gibberellins. The gibberellins will, however, not cause short-day plants to bloom when the days are long, and so are not the hypothetical hormone "florigen."

Gibberellins also promote the expansion of young leaves, may cause the breaking of dormancy of buds or seeds, substitute for the light required by some seeds before they will germinate, and promote fruit setting. One of the few things auxins and gibberellins have in common is that both may promote artificial parthenocarpy.

Synthetic Growth Substances. In addition to the naturally-occurring plant hormones there are numerous synthetic growth substances that have hormone-type effects on plant growth though they cannot strictly be called hormones since they are not naturally produced. One large group includes a variety of compounds that have effects similar to those of the auxins, although they may be more effective in some respects (as in the rooting of cuttings) and less effective in others (such as stem bending) or may fail to produce some auxin effects entirely. Some are indole compounds other than indole-3-acetic acid, such as indole-3-butyric acid. Others are naphthalene derivatives, such as alpha-naphthaleneacetic acid, or phenoxy derivatives such as 2,4-dichlorophenoxyacetic acid (2,4-D) and 2,4-5-trichlorophenoxyacetic acid (2,4,5-T). These phenoxy compounds are of great practical interest and importance since at higher concentrations than those that produce hormone-like effects they act as selective herbicides. Both 2,4-D and 2,4,5-T kill broad-leaved plants but do not kill grasses, so they are useful in controlling weeds in fields of grasses such as corn, wheat, oats, or rye, or in lawns. The 2,4,5-T is more effective against woody weeds such as poison ivy and Japanese honeysuckle than is 2,4-D.

Several different kinds of synthetic growth inhibitors are also known. These do not kill plants at the concentrations used, but stop their growth. Among them are maleic hydrazide (MH), Amo-1618 and related compounds, and chlorocholine chloride (CCC) and related compounds. MH has had more extensive practical application than the others, being used to control the growth of grass (especially along highways), to control the growth of tobacco suckers, to prevent the sprouting of such plants as onions, potatoes, and nursery stock in storage, and to reduce the labor of pruning hedges. However,

CCC has some advantages and may be used more extensively in the future, particularly in the production of compact potted plants.

There are also naturally occurring growth inhibitors, but most of these have not been isolated and identified chemically and are much less well understood than auxins and gibberellins. At least some of them are anti-auxins or anti-gibberellins, while others are specifically germination inhibitors.

Practical Uses of Growth Substances. The practical used of herbicides and growth inhibitors has already been mentioned, but it may be added that 2,4-D is commonly used in place of auxins for such things as artificial parthenocarpy and rooting of cuttings because it is much cheaper, and it is also used in much lower concentrations. 2,4-D is also used to hasten the ripening of some fruits, particularly bananas. Maleic hydrazide is quite inexpensive and can be obtained at farm and garden shops, but the other growth inhibitors are somewhat more expensive and harder to find, although Amo-1618 is available under trade names such as "Plant Tranquilizer." Powders containing auxins are available at most garden shops under trade names such as Hormodin, Rootone, and Fruitone, and are relatively inexpensive. Gibberellin is available under trade names such as Giberel. The home gardener as well as the commercial horticulturist is, then, able to secure various plant growth substances and to experiment with them and use them. The most extensive practical use of auxins is in the rooting of cuttings, but 2,4-D and other selective herbicides are the most widely used plant growth substances. Many millions of dollars are spent annually on plant growth substances by farmers, commercial gardeners, nurserymen, and home gardeners.

GROWTH CORRELATIONS

The growth of the various organs of a plant is generally coordinated, rather than being random or haphazard. This coordination of growth is brought about by the influence of one part on the growth of another, principally by means of either hormones or foods. Apical dominance is one example of a growth correlation. The shoot of a plant (its stems, leaves, and flowers) influences the growth of the roots through hormones such as auxin and thiamin and through the food supplied to the roots. The development of fruits from ovularies within which embryo plants are developing is another growth

Plant Growth

correlation, brought about by auxins. When fruits develop on a plant, particularly an annual plant, further production of flowers and vegetative growth are both generally inhibited. Removal of the fruits as they form can thus greatly extend the period of blooming and also the vegetative growth.

The shoot/root ratio is one index of growth correlations in plants and it may be influenced by both hormones and foods. For example, the amount of carbohydrate produced by photosynthesis in relation to the amount of nitrogen available from the soil (the carbohydrate/nitrogen ratio or C/N ratio) has marked influence on the shoot/root ratio. If the C/N ratio is high (either because a large amount of sugar is being produced by photosynthesis or because the available nitrogen is low) the roots will be very large in comparison with the shoots (a low S/R ratio). The reason is that there is much surplus carbohydrate beyond that used by the shoot and this is translocated to the roots. Since the roots have first chance at the limited nitrogen supply, they use most of it in converting the carbohydrate into amino acids and proteins and so have abundant food for growth. However, there is little nitrogen left for the shoots, and so their growth is limited by a lack of proteins. This means that still more carbohydrate is left over and translocated to the roots. On the other hand, if there is a low C/N ratio there is an abundance of nitrogen left over after the roots have used what they can, so the shoots can make much protein (they have first chance at the carbohydrates) and will grow luxuriantly. As a result, there is little carbohydrate left over to be translocated to the roots, and root growth is limited by lack of carbohydrate.

The C/N ratio has practical applications for the farmer and gardener. Crops where underground parts are used, such as potatoes, sweet potatoes, or carrots, do better if the C/N ratio is kept rather high (by not supplying too much nitrogen fertilizer), otherwise there will be luxuriant shoots but very limited growth of the roots or tubers. Amateurs sometimes make the mistake of fertilizing potatoes too much and are pleased with their lush plants, but when they harvest their crop they find only a few small tubers. If leaves, stems, fruits, or seeds are the desired part of the crop plant, as in lettuce, spinach, celery, peas, beans, or cucumbers, the C/N ratio should be low. Of course, in any event, there must be enough nitrogen available to avoid nitrogen deficiency symptoms.

REGENERATION

The capacity of an organism, an organ, or even a tissue to produce the parts of a complete organism that may be missing is known as **regeneration.** Plants and some invertebrate animals have very extensive powers of regeneration, but the vertebrate animals have only limited capacity for regeneration, generally only extending to the healing of wounds. The formation of adventitious roots by stem cuttings is a good example of regeneration, the result being a complete plant. Roots have more limited capacities for regenerating stems, but if the stems of many species are removed the roots will form adventitious buds that grow into aerial shoots and so regenerate a complete plant. This often occurs when trees are cut down, the "water sprouts" from the roots and stumps often growing into dense thickets if not removed. To prevent this trees are sometimes girdled by removing a ring of bark from the entire circumference of the trunk a year or two before they are to be cut down. Since food moves from the leaves to the roots through the phloem of the bark, the result is to kill the roots by starvation when the accumulated food in them has all been consumed. After the death of the roots the top of the tree dies from lack of water.

Leaves have quite variable capacities for regeneration. Those of some species such as begonias and African violets can produce both adventitious roots and adventitious buds when detached from the plant and so can regenerate a complete plant. The leaves of some other species can form adventitious roots but no buds, while the leaves of many species can develop neither buds nor roots. Leaves, stems, or roots that can regenerate all the missing organs of a complete plant can be used in the vegetative propagation of plants.

DORMANCY

In the autumn the buds of many perennial plants, particularly those that grow in temperate and frigid climates, become dormant, *i.e.,* they will not grow into branches even if all the environmental factors such as temperature are suitable for growth. Only after dormancy has been broken can the buds grow. Dormancy is apparently brought about primarily by the short days of autumn, but is broken only after the buds have been exposed to some period of cold weather. If a shrub or small tree that has become dormant is moved into the greenhouse during the winter its buds will not start growing in the spring. The dormant buds of many species are resistant to freezing injury, whereas the active buds are readily frozen and killed. Dormancy is apparently the result of production of a natural growth inhibitor which is destroyed by a sufficient period of low temperature. Dormancy may also be broken by various chemicals such as gibberellins, ether, chloroform, ethylene chlorohydrin, and ethylene dichloride.

The seeds of many, but not all, species of plants may also become dormant. One kind of seed dormancy is a result of the presence of germination inhibitors in the seeds. The germination inhibitors may be produced within the seeds or may be produced in the fruits or even the leaves of the plant and diffuse into the seeds. This type of seed dormancy is similar to bud dormancy and is commonly broken either by low temperatures or the gradual leaching of the inhibitors from the seeds by rains. Tomatoes contain germination inhibitors that produce seed dormancy.

Other types of seed dormancy result from seed coats that are impermeable to water or oxygen or are simply so hard and thick that the embryo plant cannot break them open as the seed imbibes water. In nature such seed coat dormancy is generally broken by the gradual decay of the seed coats or their cracking by alternate freezing and thawing. Seed coat dormancy can also be broken by treating the seeds with strong acids for short periods of time or by scarifying the seed coats by shaking seeds mixed with sharp sand. Clover and vetch are examples of seeds that have seed coat dormancy and must be scarified before being sold for planting.

Both bud and seed dormancy have considerable survival value, since they keep the buds or seeds from growing in the autumn when the young branches or plants would be subject to killing by freezing.

GROWTH PERIODICITY

Plants generally continue growing as long as they are alive, in contrast with most animals, but the growth is not necessarily continuous or uniform in rate. Thus, there is a definite growth periodicity, particularly in perennial or biennial plants. Trees and shrubs generally do not grow at all during the winter when their buds are dormant. In the spring, after dormancy has been broken and the environmental factors such as temperature, day length, and water availability have become favorable, growth generally occurs at a rapid rate for a period of a month or so and then begins to slow down. There is frequently little or no growth during the summer, particularly during the late summer. This seasonal growth periodicity is reflected in the annual rings of the wood. However, even during the spring or the summer growth is frequently not steady and there may be cycles of slower and more rapid growth even within a season.

There is also a daily growth periodicity in most plants, growth usually being more rapid during the night than during the day. This is primarily a result of the inhibition of growth by light, but may also be influenced by other factors such as the generally greater turgidity of plants at night. Plants that are wilted or are beginning to wilt hardly grow at all.

Chapter 9

ENVIRONMENT AND PLANT GROWTH

As every farmer and gardener knows, the environment of plants has marked influences on their growth and the successful cultivation of plants really involves only two main sets of factors: securing plants of the best possible heredity and then providing them with the best possible environment. The environment of a plant consists of all the other organisms that in any way affect the plant (its **biological** environment) and of the various substances and types of energy that the plant is exposed to (its **physical** environment). The biological environment will be considered in the next chapter, so we shall consider only the physical environment here.

The principal types of energy that influence plant processes and plant growth are light and other radiation, heat (temperature), and gravity. Electricity, magnetism, and mechanical movements (such as the bending and twisting of trees by the wind) may also have important influences on plant growth, but relatively little is known about them and they will not be discussed here. Ordinary sound waves have no apparent influences on plants, but high frequency sound waves that we cannot hear (ultrasonics) can disrupt and kill organisms, particularly unicellular plants and animals. The principal substances in the environment of plants are water, the gases of the air, mineral salts, acids and bases, and the pulverized rock particles and organic substances that make up the soil. The influence of mineral salts on plant growth is marked and has already been discussed in Chapter 6.

WATER

The importance of water in the life of plants has been noted in Chapter 5, but it should be emphasized here that an adequate supply of water is essential for plant growth and that wilted plants grow little if at all. However, most plants do not grow well if the soil is poorly drained and waterlogged. The adverse effects result from the displacement of air from the soil by the water and the consequent reduction in respiration, and even suffocation of the roots because of a lack of oxygen. The short, yellowish corn plants in a low and waterlogged part of a field provide a striking example.

Perhaps the most outstanding example of the influence of water on the pattern of plant development, as contrasted with the rate of growth, is on the kinds of leaves produced by plants that grow partly below the water in a pond or lake and partly in the air. The submerged leaves of a number of such species are generally highly divided or lobed, while the leaves in the air are more nearly entire (Fig. 70). However, this difference in development is probably not a direct effect of the water but rather a result of the influence of the water on other factors such as temperature, light intensity, oxygen concentration, or perhaps even the length of the daily light period (since the reduction in light intensity under water means that it will get dark a little earlier in the evening and light a little later in the morning than in the air above).

SOIL

Aside from its water, air, and mineral salt contents, soil influences plant growth principally through its structure and acidity. A loose sandy or loam soil, or a soil with a high content of humus, provides a much easier medium through which roots can grow than a heavy clay soil which is hard when dry and sticky when wet. The thickness of the soil layer above the underlying rock is also an important factor in plant growth. Plants with deep root systems can-

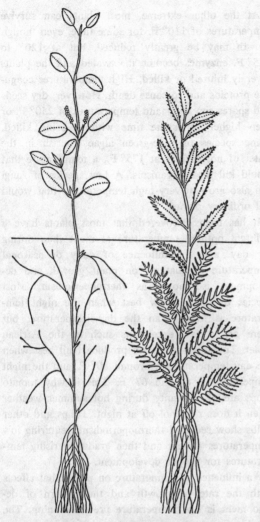

Fig. 70. Two species of amphibious plants showing difference in shapes of the submerged and aerial leaves.

not thrive in a thin soil, and the soil is likely to be waterlogged frequently.

The acidity of a soil has marked influences on plant growth. The degree of acidity is usually measured by the *p*H scale, which extends from 0 to 14. Each *p*H unit is ten times less acid than the next lower one, so *p*H 6 is 10 times less acid than *p*H 5, 100 times less acid than *p*H 4, and 1000 times less acid than *p*H 3. Neutrality occurs at *p*H 7, anything below this being acid and everything above it alkaline. Most plants grow best when the *p*H of the soil is 6 to 7, but certain species such as cranberries, azaleas, camellias, and some hollies grow well only when the soil is rather acid, *i.e.*, around *p*H 4 to 5. Soils are generally alkaline only in arid regions, where the *p*H may reach 8 or 9. Most plants grow poorly in such alkaline soils, but a few species native to such arid regions can flourish at these high *p*H levels. At extreme ranges of *p*H plants may be quite stunted and appear to have mineral deficiencies. This may actually be the case, since one adverse effect of low or high acidity is the tying up of various mineral elements in insoluble forms unavailable to plants. The principal practical means of increasing the *p*H of acid soils is to add lime to the soil, commonly in the form of finely ground limestone. Several substances such as aluminum sulfate can be used to reduce the *p*H of a soil if necessary, *i.e.*, to make it more acid.

GASES OF THE AIR

Since plants use oxygen in respiration and carbon dioxide in photosynthesis, it is essential that adequate quantities of these gases be available in the air. As we have noted, oxygen is practically never deficient enough in the air to limit respiration, but the lack of air in waterlogged soils may hamper plant growth. Carbon dioxide is frequently a limiting factor in photosynthesis, and so may in turn be a limiting factor in plant growth. Of greater concern to those who raise plants or who are interested in the natural vegetation are the toxic gases added to the air by certain types of manufacturing processes and by the exhaust fumes of automobiles. The commonest of these is sulfur dioxide, but there are many others. Smog has very detrimental effects on the growth of plants, and may in some cases kill them.

The rate of air flow as well as the composition of the air may influence plant growth. Continuous strong winds from one direction, as on a high mountain, greatly reduces the rate of tree growth and results in trees growing at a marked angle, often with branches only on one side. Similar bent and distorted trees and shrubs are often seen along a seacoast, but here an important factor restricting plant growth is the salt spray carried by the wind. The high salt content of the air may plasmolyze and kill the leaves and young branches on the side toward the wind. Winds also influence plant growth somewhat more indirectly by causing soil erosion and dust storms. Of course, very high winds may destroy plants, tornadoes by splintering them and hurricanes by uprooting them.

GRAVITY

The most important influence of gravity on plant growth is geotropism, which has already been discussed, but gravity also has several other effects. The underside of a horizontal tree branch may grow much more rapidly than the upper side, as evidenced by the greater thickness of the rings on the lower side. Gravity may also influence leaf size and shape. This is most evident in trees with opposite leaves, one pair being attached horizontally and the next vertically, for instance, maples. In such leaves the ones hanging down are larger than the ones extending up, while in the horizontal leaves the upper halves are smaller than the lower halves. Such effects of gravity, like geotropism, probably result from the influence of gravity on the distribution of auxins.

TEMPERATURE

Just about every life process of plants is influenced by temperature, so we can expect temperature to have rather marked influences on plant growth. Most species of plants grow better at moderate temperatures of 70° to 80° F. than at higher temperatures, though different species have differing optimal temperatures. There are several reasons why high temperatures may be unfavorable for growth, one being the high rate of transpiration and another the low photosynthetic/respiratory ratio.

Extremely high or extremely low temperatures may be injurious or even lethal to plants. While some species are able to survive through the lowest temperatures found on earth if they are dormant, many other species are killed by temperatures just below freezing. The nature of resistance to freezing injury is not well understood, but it is possible to harden plants so their resistance to freezing will be increased. This is a standard horticultural practice. Thus, young cabbage plants transplanted out of doors directly from a warm greenhouse might be killed by a mild freeze, whereas if the plants are first hardened by being kept at temperatures just above freezing for a week or so they can survive mild freezing. Freezing injury involves several factors such as the disruption of protoplasm by ice crystals and the desiccation of the protoplasm as the water freezes into ice.

At the other extreme, most plants can survive temperatures of 110° F. for some time, even though growth may be greatly reduced, but at 120° to 125° F. enzymes become inactivated and the plants severely injured or killed. High temperatures coagulate proteins and so cause death. However, dry seeds and spores can withstand temperatures of 250° F. or even higher for some time without being killed. Some species of blue-green algae flourish in the water of hot springs at 175° F., a temperature that would kill most organisms. A few species of fungi can also grow at very high temperatures that would kill ordinary plants.

It has been discovered that most plants have a different optimal temperature at night than during the day, and the influence of daily or seasonal temperature fluctuations on plant growth and development is known as **thermoperiodism.** Most species of plants grow best when the night temperature is lower than the day temperature, but there are a few exceptions such as the African violet. Tomatoes grow and produce fruit best when the day temperature is around 80° F. and the night temperature is around 67° F. This is why tomato crops are often scanty during hot summer weather when it does not cool off at night. Tulips and other bulbs show seasonal thermoperiodism, requiring low temperatures at first and then gradually rising temperatures for proper development.

An influence of temperature on plants that affects both the rate of growth and the pattern of development is **low temperature preconditioning.** The breaking of bud and seed dormancy by low temperatures may be considered as low temperature preconditioning, but the plants most subject to low temperature preconditioning are biennials. During the first year of its life a biennial plant such as celery, cabbage, carrot, henbane, or winter wheat has stems with very short internodes and does not bloom. The leaves on the short stems of many biennial species are flat against the ground, forming a rosette. The low temperature preconditioning during the winter is followed in the spring by rapid and extensive elongation of the internodes (bolting) and the blooming of the plant. Once seeds and fruits have matured the plant dies. Bolting and blooming can be induced the first year, rather than the second, by subjecting the young plants to a week or so of low temperatures, just above freezing. The low temperature treatment can be applied even to soaked seeds, a process known as **vernalization.**

Vernalization has been used to some practical extent with winter wheat. Spring wheat is an annual, and does not require low temperature preconditioning. We have noted earlier that gibberellins can substitute for low temperature preconditioning in bringing about the bolting and blooming of biennials.

LIGHT

Fewer plant processes are influenced by light than by temperature, but the effects of light on plant growth and development are more diverse and striking than those of temperature. For each process influenced by light there is a pigment, a substance that absorbs certain wave lengths of light more than others and so has a characteristic color. The best known of these pigments are the chlorophylls, which absorb the light energy used in photosynthesis. Because it is the source of a plant's food, photosynthesis plays an important role in growth. Only when photosynthesis produces more food than is used in respiration is there food left over that can be used in assimilation and accumulation.

However, photosynthesis is only one of a number of plant processes in which light plays a role. We have already noted that light influences auxin concentration, and so stem elongation and phototropic bending. Stems grow more in the dark than in the light, and the darker side of a stem grows faster than the more brightly lighted side. The pigment concerned in phototropism is apparently riboflavin (a yellow vitamin of the B group), although some botanists have evidence indicating that the carotenoids may be the effective pigments.

By causing the opening of stomates light has an influence on transpiration and also the exchange of gases in photosynthesis and respiration. Light is required for the synthesis of chlorophyll in most angiosperm plants, a role quite different from that in photosynthesis. Consequently, plants growing in the dark lack chlorophyll and have a pale yellow color. Light is also essential for the formation of some anthocyanins, the red pigments of leaves, flowers, and some fruits. However, the role of the light absorbed by the anthocyanins, if any, is not known.

When a plant is growing in the dark it not only has tall, spindly stems and lacks chlorophyll, but also has small leaves that do not unfold or increase in size and, in some plants like beans, hypocotyls that do not straighten out from their hooked shape. This combination of characteristics is referred to as **etiolation.** A plant may be partially etiolated even when growing in dim light, particularly with respect to stem length.

The expansion of the leaves and the opening of the hypocotyl hooks are both dependent on light absorbed by a pigment called **phytochrome.** This pigment has a bluish green color but there is so little present in plants that it imparts no visible color to them. It exists in two forms convertible into one another: one absorbs light in the red part of the spectrum and the other absorbs far-red rays just at the limits of visibility between red light and infrared rays. When the red-absorbing form absorbs light it is converted into the far-red absorbing form, and vice versa. Red light promotes leaf expansion and hypocotyl hook opening, while far-red counteracts the effects of the red. If an experimental plant kept in the dark except for brief exposures to red or far-red light is exposed to red light, leaf expansion and hypocotyl hook opening will occur, but if far-red follows the red they will not occur. A second exposure to red light will promote them and a second exposure to far-red will cancel the effect. The last exposure is always the effective one. When white light is used the red effect predominates and leaf expansion and hypocotyl hook opening occur.

A good many other plant processes involving phytochrome and the reversible red, far-red reactions have been identified. Some seeds, such as those of tobacco, peppergrass, and Grand Rapids lettuce, will not germinate unless they are exposed to light for at least a brief period after they have imbibed water. Exposure to red light promotes germination, while subsequent exposure to far-red cancels the effect of the red. As with leaf expansion and hypocotyl hook opening, the last treatment in a series is the effective one. The formation of anthocyanin in the skin of pink tomatoes also appears to be a red, far-red reaction involving phytochrome.

Photoperiodism, the influence of day length on plants, also is a red, far-red process and again phytochrome is the effective pigment. The length of the daily light period has a variety of influences on plants. Most plants grow larger under long days than under short days, and this effect is quite independent of the longer time for food manufacture by photosynthesis since the day length can be extended by weak artificial light that permits little photosynthesis. Short days promote leaf abscission and dormancy, and tuber formation is also favored by short days. However, onion bulbs form only under long days. The small plantlets in the notches of Bryophyllum (air plant) leaves develop only under long days. The most striking and widespread effects of day length on plants, however, are on the initiation of flowers. Some plants such as chrysanthemums, poinsettia, cocklebur, cosmos, and some varieties of tobacco and soybeans are short-day plants and bloom only when the day length is below a certain critical length. Other species such as spinach, radish, dill, rose mallow, larkspur, phlox, and clover are long-day plants, and bloom only when the day length is above a certain critical length. The critical photoperiod varies from species to species but in general is around fourteen hours of light. A third class of plants (day neutral plants) includes those that produce flowers under either long or short days, such as tomatoes, corn, snapdragon, garden beans, peas, cucumbers, and most varieties of tobacco. Although corn forms flowers under both long and short days, there are commonly abnormalities such as small ears in the tassels under short days. Day length may cause a change from staminate to pistillate flowers or vice versa in other species, too. Rudbeckia and a number of other species require long days, not only for flower formation, but also for elongation of the stem internodes, rosettes being formed under short days.

Photoperiodism, rather than temperature, is the principal factor controlling the seasonal blooming of plants, but temperature may influence the critical day length or the plant's photoperiodic sensitivity. For example, orange flare cosmos is short day at low temperatures but day neutral at higher temperatures. The gradual change in day length from the shortest day in December to the longest day in June and then back again is uniform and predictable, in contrast with wide variations in temperature from year to year on a certain day, and so provides an effective timing device. Day length varies with latitude as well as with season. At the equator day length is the same all year long and as one goes from the equator to the poles the seasonal changes in day length get more and more marked, culminating at the poles in continuous light in the summer and continuous darkness in winter.

For experimental or practical purposes a short day may be made into a long one simply by means of ordinary electric bulbs. As little as a fraction of a foot-candle of light is sufficient, though the light of the full moon is not quite bright enough to be photoperiodically effective. Long days can be converted into short days simply by placing the plants in the dark around four or five o'clock and keeping them dark until eight or nine in the morning. Greenhouse operators have made extensive use of both lights and dark cabinets to bring plants into bloom out of season. This practical application of photoperiodism has brought great economic returns, especially with seasonal plants like Easter lilies and poinsettias that are in little demand after a certain time.

It has been found that if the long night accompanying a short day is interrupted by a half hour or so of bright light near the middle of the night short-day plants will not bloom and long-day ones will, even though the total length of the light periods is still well below the critical. This suggests that important processes related to photoperiodism occur slowly at night and are interrupted by light, and suggests that short-day plants should perhaps really be called "long-night plants," while long-day plants should be called "short-night plants." Breaking a long day with a period of darkness near its middle does not change the photoperiodic response of plants. Red light is effective in breaking the long dark periods, but far-red following the red destroys its effect.

Little is known about the internal processes involved in photoperiodism beyond the fact that phytochrome is the light-absorbing pigment, and that the leaves of the plant are the organs sensitive to the light. There is evidence that processes in the leaves occurring when no light is being absorbed by phytochrome result in the production of a substance called **florigen** that is translocated to the very young buds and causes them to develop into flower buds rather than leaf buds. However, florigen has never been isolated. It can pass from one plant to another if their stems are grafted together, causing the second plant to bloom even though it has never received a suitable photoperiod. The reason for the difference in responses of long-day, short-day, and day neutral plants to day length is not known.

It is likely that still other red, far-red reactions mediated by phytochrome will be discovered. Although phytochrome has been identified only in recent years it appears to be a pigment of very great importance in plants.

OTHER RADIATION

The light waves, that is, the visible wave lengths from violet at one end of the visible spectrum to red at the other, constitute only a small portion of the spectrum of electromagnetic radiation. Longer than the red waves are the infrared or heat rays and the electrical or radio waves. Shorter than the violet waves are the ultraviolet, X rays, gamma rays, and cosmic rays. Infrared radiation influences plants mostly, if not entirely, by raising their temperature, but heat is dissipated quite rapidly from plants. As far as is known, radio waves have little if any influence on plants, although certain bands of short radio waves of high intensity have been found to have some effects on animals and microorganisms and may prove to influence plants, too. These radio waves are not part of the natural environment.

The radiation with wave lengths shorter than visible light, however, may have quite marked influences on organisms. Certain wave lengths of ultra-violet are lethal to microorganisms and the superficial tissues of larger organisms, but fortunately they are mostly filtered from sunlight by the atmosphere. The X rays, gamma rays, some wave lengths of ultraviolet, and perhaps also cosmic rays cause the ionization of molecules that ordinarily do not form ions, and so are called **ionizing radiation**. X rays are not a part of the natural environment and gamma rays are produced only by certain radioactive substances. High-speed particles such as beta particles (electrons), protons, and alpha particles (helium nuclei) that are emitted by radioactive substances also cause ionization and are sometimes included with ionizing radiation, although they are really not true radiation. The influence of ionizing radiation in increasing the mutation rate is now well known, and will be discussed in Chapter 13. Mutations rarely show up in the individual in which they occur, but ionizing radiation may also have immediate adverse effects on the structure and processes of plants, particularly when the intensity is high. The organs of a plant developed after exposure to ionizing radiation may be quite distorted and abnormal.

Chapter 10

BIOLOGICAL INTERACTIONS

The biological environment of a plant may have effects on its processes, growth, and development that are just as important, extensive, and striking as those of the physical environment. If the interactions between one species and another do not involve the securing of food from one by the other, the interactions are referred to as **social,** while those involving food are known as **nutritive.** We shall consider the principal types of each class of interactions.

SOCIAL INTERACTIONS

The growth of a vine over a tree is one type of social interaction. Although neither one secures food from the other the vine may influence the tree by shading its leaves and perhaps by constricting its trunk if the vine is large and woody and coils around the trunk, as in the wisteria vine. **Epiphytes,** plants that grow on other plants and have no roots in the soil, also interact socially with the trees supporting them. Unlike parasites, they do not get food or water from the tree but they may shade the leaves of the tree and so interfere with its photosynthesis. Many tropical orchids are epiphytes, as is the Spanish moss of the southeastern states. Both vines and epiphytes secure support and better exposure to the light from the trees on which they are growing. The trees of a forest interact socially with the smaller plants and the animals of the forest, largely by altering their physical environment. For example, they reduce the light intensity and temperature, increase the humidity of the air, and add organic matter to the soil as their leaves fall and decay. The relations between man and the ornamental plants (or other non-food plants) that he cultivates may be regarded as social interactions.

One important type of social interaction is antibiosis, *i.e.,* the production of antibiotics that inhibit the growth of certain other organisms in the vicinity of the antibiotic-producing organism. The result is likely to be reduced competition for food, water, minerals, space, and other essential environmental factors. The best-known antibiotics are those such as penicillin that are produced by fungi and inhibit the growth of various species of bacteria, but it is now known that plants of many different types such as algae and seed plants also produce antibiotics. For example, walnut trees and the desert shrub encelia produce antibiotics that keep other species of seed plants from growing near them. Some antibiotics apparently prevent the growth of parasites in plants producing them. Antibiotics may be an important factor in determining the composition of plant communities and in the succession of different species in a developing plant community.

NUTRITIVE INTERACTIONS

One obvious type of nutritive interaction is the eating of one organism by another. Animals are, of course, the organisms usually doing the eating, but carnivorous plants such as Venus's-flytrap, the sundews, the bladderworts, and the pitcher plants reverse the usual situation of animals consuming plants. Some animals live directly on plants and are called **herbivorous** animals. Among the herbivorous animals are many species of insects and some other invertebrates, and many kinds of birds. **Carnivorous** animals live on herbivorous animals, or more rarely on other carnivorous animals. Some animals, such as man, various birds, pigs, and rats are **omnivorous,** eating both plants and animals.

A second type of nutritive interaction is **parasitism.** A parasite is an organism that lives in or on another organism (the **host)** and obtains its food from it.

Although many parasites cause disease of their hosts, and so are referred to as **pathogens,** parasites are not always pathogens and may live in or on the hosts for long periods of time without causing any serious disorders. Thus, typhoid carriers are parasitized by typhoid bacteria without any of the symptoms of typhoid fever. Plants have many parasites, particularly various species of fungi, bacteria, and nematode worms. Some insects, such as the scale insects, are parasites on plants. It is sometimes rather difficult to draw a sharp line between animals that are parasitic on plants and those that eat plants and so fall in the first group, particularly when dealing with insects, but in general a parasite lives in or on the plant most of its life rather than moving about from plant to plant. Parasites are always much smaller than their hosts.

Fungi that are particularly destructive parasites of plants include the various rusts and smuts, Dutch elm disease, downy and powdery mildews, the blue mold of tobacco, potato blight, anthracnose, black and brown rots, and damping off of seedlings. Bacteria cause a variety of blights and wilts, but in general are not the cause of as many plant diseases as fungi. A few seed plants are parasites on other plants, perhaps the most widespread and destructive being dodder. There are many species of dodder that are parasitic on numerous species of plants. Soon after the young dodder plants become attached to their hosts the roots die and thereafter the dodder is dependent on the host plant for water and minerals as well as food. Dodder is a vine with very small leaves and practically no chlorophyll.

Mistletoe is at least a partial parasite, although its leaves contain chlorophyll and it probably makes part of its own food.

A serious parasite of corn and other grasses, called witchweed, has only recently been introduced into the Carolinas from Africa and efforts are being made to keep it from spreading. It has green leaves and makes some of its own food, but is attached to the roots of the host plants and does considerable damage to them. It has orange flowers and produces numerous very small seeds. Other parasitic seed plants include the beechdrops that grow on the roots of trees but apparently cause little damage to them. Nematode worms are probably the most serious animal parasites of plants and cause extensive crop losses. They attack the roots of plants, and remain in the soil from year to year. When plants are growing poorly for no evident reason they are frequently parasitized by nematodes. Soil fumigants for killing nematodes are available.

The **alternate host** parasites constitute an interesting type. These parasites attack two different hosts of entirely different species and can complete part of their life cycle only on one of the hosts and part only on the other host. A well-known example is the malarial parasite, which has man and a mosquito as its hosts. The rusts are the principal alternate host parasites of plants. Among the more important ones are the black stem rust of wheat, with wild barberry bushes as the alternate host, the cedar apple rust of apple trees and cedar trees, and the white pine blister rust which has currant or gooseberry bushes as the alternate host. Alternate host parasites can be controlled by destroying the less valuable host. In the examples given these would be the barberry, the cedar trees, and the currant and gooseberry bushes.

Many plants are parasites of humans and other animals. Among the bacterial parasites are those causing typhoid fever, tuberculosis, pneumonia, diphtheria, and the streptococcus and staphylococcus bacteria that cause boils and other infections. Many different fungi are parasitic on man, including those that cause ringworm, athlete's foot, and a variety of infections of the ears and lungs. Fungi are also parasitic on other animals, on lower plants such as algae, and even on other fungi.

Some parasites of plants cause the host tissues to develop into abnormal and often elaborately shaped structures known as galls. One group of wasps is a particularly important cause of plant galls, and they parasitize oak trees more than any other kind. The wasps lay their eggs in the leaves or stems and as the eggs develop into the larvae the tissues of the plant surrounding them grow into a gall. The larvae use the gall tissues as food. Each species of gall wasp parasitizes a certain species of plant, the wasps being able to distinguish between the different species of oaks. Galls are also caused by other kinds of insects and by other organisms such as fungi and bacteria. Although galls are generally conspicuous and striking structures, gall organisms are not serious pathogens and the plants are usually not damaged greatly by their presence.

In contrast with the two types of nutritive interactions we have discussed is the third type—**symbiosis** or **mutualism.** In a symbiotic relationship one organism gets food from the other, but the host organism also gains a benefit from the relationship.

For example, bees get their food (pollen and nectar) from the flowers of plants they visit but in turn the bees pollinate the flowers. Ants get food from plant lice (aphids) by stroking them and then consuming the drop of sweet honeydew the aphids secrete, but the ants take care of the aphids, moving them from one plant to another and taking them into their nests during the winter. The relationship is much the same as that between man and the cows that provide milk.

The interaction between man and his crop plants is really symbiotic, since most of them are so highly specialized that they could no longer survive in nature without man's care. Similarly, there is one species of ant that cultivates a fungus and uses it for food, and it is possible that this fungus would not thrive without the ants' care.

The relationship between legumes and the nitrogen fixing bacteria that live in the nodules on their roots is symbiotic. The bacteria secure food from the legumes while the legumes secure nitrogen fixed by the bacteria. Fungi live in or on the roots of a good many plants, particularly forest trees, a number of the different forest mushrooms being the reproductive structures of these fungi. The fungus-root complexes are known as **mycorhiza** and constitute a symbiotic relationship. The fungi get food from the trees while in turn they play a role in the absorption of water and minerals by the trees. Trees with mycorhiza do not thrive without them and generally have few or no root hairs on their roots.

Lichens, the compound organisms consisting of unicellular algae living within the hyphae of a fungus, are frequently considered to be symbiotic. The fungus gets food from the algae, and presumably the fungus protects the algae from drying out and permits them to grow in much drier habitats than would be possible otherwise. However, some authorities believe that the algae do not gain any benefits from the relationship, at least in some lichens, and so consider the interaction to be parasitism rather than symbiosis.

FOOD PYRAMIDS AND ENERGY FLOW

We have noted several times that all living things (except the chemosynthetic bacteria) depend on photosynthetic plants for their food. In any community of living organisms, then, photosynthetic plants constitute the broad base of the food pyramid and may be regarded as the key organisms in the community. These green plants are frequently referred to as the **producers,** since they make enough food not only for their own respiration and assimilation but also for the respiration and assimilation of all the other organisms of the community. The herbivorous animals that live on the plants directly are referred to as the **first-level consumers,** the carnivorous animals as **second-level consumers,** and the carnivorous animals that consume other carnivorous animals as **third-** and **subsequent-level consumers.** In land communities there are few animals at these higher levels of consumption, but in the ocean there are many levels of consumer organisms. In addition to animal consumers that eat plants or other animals there are the parasites that constitute consumers at the first level (if they live on green plants) or at higher levels if they live on animals or non-green plants.

In a forest community the trees, shrubs, and herbaceous plants constitute the producer organisms. The first-level consumers include animals such as rabbits, squirrels, chipmunks, and a variety of birds and insects, and also the numerous parasites on the plants. The second-level consumers include animals such as wolves and foxes, birds such as owls, and the parasites of the herbivorous animals. In an ocean community the principal producers are microscopic algae, and the first-level consumers are tiny crustaceans and other animals. These are in turn consumed by larger crustaceans such as shrimp or by small fish or other small animals, the many consumer levels eventually leading to the largest fish or other marine animals. In general, the size of the organisms keeps increasing as one goes along the chain of eaten and eater and the number of organisms decreases. However, in a parasite-host chain the size of the parasites keeps decreasing and the number of individuals keeps increasing as one progresses from level to level.

Whatever may be the situation in regard to sizes of organisms, there is always a smaller total bulk of living organisms at any level than at the one below it, by a factor of about ten. Thus, for every thousand pounds of green plants in a community there would be only about one hundred pounds of first-level consumers, ten pounds of second-level consumers, and one pound of third-level consumers. This food pyramid is a biological necessity, for if the consumers at any level were anything near as abundant as the organisms at the next lower level on which they depend for food, these organisms would be entirely or largely consumed and starvation would result in the collapse of the food pyramid. It is not just a matter of chance that in any balanced natural community the plants are by far the most abundant as well as the most conspicuous organisms.

In addition to the plants and animals in the main food pyramid of a community there are **decomposer organisms.** These live on the remains of dead plants or animals or on animal wastes rather than on living organisms, and include many different saprophytic fungi and bacteria that cause the decay of plant and animal remains. There are also some animals such as termites and other wood-destroying insects, scavenger animals including birds such as buzzards, and a variety of insects that consume dead animals. The decomposer organisms, particularly those that cause the decay of plant and animal remains, play essential roles in a community by breaking down the remains into simple inorganic substances such as water, carbon dioxide, and mineral salts that can then be used over again by plants and so started through the food chain once more. If there were no decay or other decomposition of the plant and animal remains, the essential chemical elements would eventually all become tied up in a vast grave-yard that would clutter up the face of the earth and life would eventually become impossible. Any particular atom of carbon, hydrogen, oxygen, nitrogen, or other essential element has probably been circulated through the food chains of communities many times, from the environment to the producer plants, through the various levels of consumers, and through the decomposers back to the environment. A little later on we shall consider in some detail the cycles of two of the more important elements—carbon and nitrogen.

While matter circulates endlessly into, through, and out of the organisms of a community, the flow of energy through the community is a one-way affair. Energy, which is so important in the life of any organism, enters a community when the plants carry on photosynthesis, the energy from sunlight being converted into the chemical bond energy of sugars and then of the substances made from the sugars. Plants use some of this energy in their own respiration. Each level of consumer organism and also the decomposer organisms also use energy from their foods in carrying on various life processes after they have made the energy available by respiration. At each successive level of the food pyramid there

is not only less food, but also less energy available, since the energy is obtained from the foods. Some of the energy released from foods during respiration is immediately converted into heat, while most of it is incorporated in the chemical bonds of ATP and can be used in doing useful work. Eventually, however, all the energy that was present in foods becomes converted to heat and is dissipated, so continuation of life in the community depends on continued trapping of energy from the sun by photosynthesis.

THE CARBON AND NITROGEN CYCLES

As representatives of the cycles through which the various essential chemical elements pass we shall choose the carbon and nitrogen cycles. The carbon cycle is somewhat simpler (Fig. 71). Carbon enters a biological community largely as carbon dioxide that is used in photosynthesis, and so is incorporated in sugars and in turn in other foods and the various organic compounds of the plants. When the plants use food in respiration the carbon is converted back to carbon dioxide and diffuses into the air. Similarly, the various levels of consumer organisms and also the decomposer organisms add carbon dioxide to the air as a by-product of their respiration. All this carbon dioxide, plus that produced by combustion of plant or animal remains, is once more available for use in photosynthesis and so the carbon is channeled once more through the living world.

The great reservoir of nitrogen is the nitrogen gas of the atmosphere. Over 79 per cent of the air is nitrogen, but most organisms are unable to use it. Only a few species of bacteria and algae have the ability to fix this atmospheric nitrogen, *i.e.,* to convert it into compounds such as ammonia that can be used by green plants (Fig. 72). All the nitrogen-fixing algae and some of the nitrogen-fixing bacteria are free-living in the soils or waters of the earth. The nitrogen they fix is partially used by themselves and is partially available to green plants. One group of nitrogen-fixing bacteria lives symbiotically in the roots of leguminous plants such as beans, clover, alfalfa, and peas. Such plants are able to use the fixed nitrogen directly and can thrive in soils deficient in nitrogen. Farmers frequently raise legumes in rotation with other crops to increase the nitrogen content of their soils without adding fertilizers.

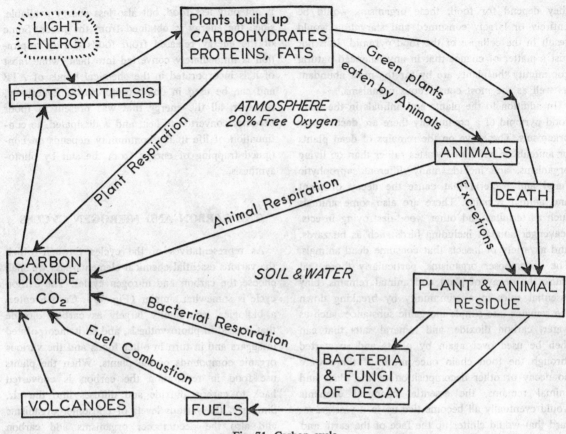

Fig. 71. Carbon cycle.

Lightning and other electrical discharges also fix nitrogen, the nitrogen compounds then being carried to the earth by rains, but the bulk of nitrogen fixation is carried on by bacteria and algae. Electrical fixation results in the production of nitrates, while biological fixation probably results in the production of ammonium compounds. Where electric power is cheap, electrical nitrogen fixation is carried on in factories, providing nitrates to be used in fertilizers, in making explosives such as TNT, and in other ways.

If green plants absorb ammonium compounds from the soil or get them from symbiotic nitrogen fixation, they can be used directly in making amino acids and then proteins or in making other compounds of which nitrogen is an essential part. If nitrates are absorbed the plants must first convert them to ammonium compounds, a reduction process that requires much energy from respiration. The amino acids and proteins of plants provide the

basic nitrogen supply of animals and of many bacteria and fungi, although some bacteria and fungi can make amino acids as plants do from sugar and ammonium compounds. Animals digest the proteins of the plants they eat and use the resulting amino acids in making their own proteins. Nitrogen is similarly transferred from one consumer level to another and to the decomposers as a constituent of proteins.

Some of the amino acids of animals are broken down and their nitrogen is excreted as urea or uric acid. Plants can use these as nitrogen sources, but usually they are converted to ammonia by bacteria. Decay bacteria and fungi may also break down amino acids and produce ammonia. The ammonia does not generally remain in the soil long, for certain bacteria (the nitrifying bacteria) convert the ammonium compounds into nitrites, and another kind of nitrifying bacteria then convert the nitrites into nitrates. These are energy-releasing oxidation

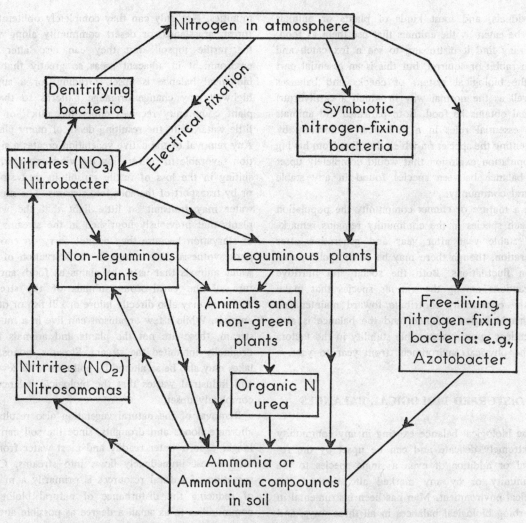

Fig. 72. Nitrogen cycle.

processes, the energy being used by the nitrifying bacteria in producing food from carbon dioxide and water by chemosynthesis. The resulting nitrates are quite stable and may remain in the soil for some time until they are absorbed by a green plant or are leached from the soil by rains. Thus, when nitrogen has once been fixed it may be used and reused.

However, the soil also contains denitrifying bacteria that convert nitrates to nitrogen gas and so the original cycle may be completed. Of course, farmers would like to provide conditions unfavorable to the denitrifying bacteria but favorable to the nitrogen-fixing, decay, and nitrifying bacteria. This is not as difficult as it might appear, since the denitrifying bacteria thrive in the absence of oxygen while the

others thrive when oxygen is abundant. Thus, by keeping a soil well drained and thus well aerated, it is possible to favor the desirable bacteria of the nitrogen cycle and to make conditions unsuitable for the denitrifying bacteria.

BIOLOGICAL CHECKS AND BALANCES

Every species of plant and animal has a much greater potential for reproduction and increase in population than is ever realized in nature. Seeds or spores may fail to land in a suitable environment, and so will die rather than grow into a mature plant. The many parasites of any species kill numerous individuals, and most kinds of plants or animals may be eaten by the animals that use them as food. We may find it distressing to see a fox catch and eat a rabbit or squirrel, but this is an essential part of the biological system of checks and balances as well as the normal way in which a carnivorous animal obtains its food. Both parasites and animals play essential roles in a biological community by preventing the species on which they live from having a population explosion that would completely upset the balance between species found in any stable natural community.

In a mature or **climax** community the population of each species in the community remains remarkably stable year after year and generation after generation, though there may be some cyclic population fluctuations. Both the social and nutritive interactions among the various species that make up the community contribute toward maintenance of this biological balance, and the balance is also dependent upon a reasonable stability in the factors of the physical environment from year to year.

DISTURBED BIOLOGICAL BALANCES

The biological balance existing in any community is extremely delicate and can be upset by the removal or addition of even a single species to the community or by any marked alteration of the physical environment. Man has been instrumental in disturbing biological balances in all three ways, and during the relatively short period of human civilization man has perhaps been more responsible for disturbed balances than any other single agent. However, natural changes in climate or topography cannot only upset a biological balance but can also completely destroy a community. For example, forests once flourished in our Southwest but a marked decrease in rainfall resulted in the death of the forest trees and the other forest species and replacement of the forest communities by the present desert communities. Submersion of coastal areas results in destruction of the land communities and the appearance of salt marsh or marine communities. Such major changes are beyond human control, but man can reduce or avoid his own disruption of biological balances, though there is little evidence that he is doing so.

The bulldozer and related machines are man's prime devices for the destruction of biological communities. Not only can they completely obliterate a forest, grasslands, or desert community along with the fertile topsoil, but they can also alter the environment in adjacent areas so greatly that the biological balance is upset. Grading for a superhighway may change drainage patterns so that a plant community receives either too much or too little water, with the resulting death of many plants. Any removal of the native vegetation creates a situation favorable to soil erosion by wind or water, resulting in the loss of fertile topsoil in dust storms or by transport of the soils into streams. The muddy water may transmit so little light that the water plants that previously flourished in the streams die of starvation because they cannot carry on enough photosynthesis. This results in the starvation of the water animals that used the plants as food, and in turn of the carnivorous animals of the stream. The mud may also directly injure or kill fish or other animals. While a few organisms can live in a muddy stream, these are not the plants and animals that originally inhabited the stream. Streams, ponds, or lakes may also be so altered by pollution with sewage and industrial wastes that the biological balance is completely upset.

Removal of the natural vegetation also results in alternate floods and droughts, since the soil can no longer absorb water readily and most water from a rain almost immediately flows into streams. Conservation of natural resources is primarily a matter of reducing the disturbance of natural biological communities to as small a degree as possible and of trying to re-establish vegetation in areas where it never should have been removed. Of course, in any civilized society there must be considerable destruction of the natural vegetation if there is to be land for farming, cities, industries, highways, and other features of civilization. However, there has been much unnecessary destruction of forests and grasslands that are valuable for timber, grazing, the prevention of erosion and the silting of streams, and as recreational areas.

Many such cleared areas are not really suited to farming because they are either too hilly or too dry, and it is in such areas that forests or grasslands should be re-established. Selective cutting of forest trees for timber along with either natural seeding or planting of trees can maintain a forest and provide a continuous source of trees for lumber, pulpwood, or other uses.

Aside from such aspects of conservation linked

with utilization of natural resources is the conservation of natural vegetation as recreational areas and as areas of biological interest for research and natural outdoor museums. As cities grow and new highways spread across the country, many valuable and attractive natural areas are being destroyed and there is a danger that little of our native vegetation will escape. Many areas should be added to our National Parks or National Forests systems or placed under the protection of organizations such as the Nature Conservancy before they are destroyed.

Biological balances can also be disturbed by less obvious ways than such outright destruction of biological communities, principally by either adding or removing one or more species of plant or animal. The introduction of plants or animals from other countries or other areas has frequently caused much trouble. Many of our bad weeds, serious insect pests such as the Japanese beetle, harmful fungus parasites such as those causing the Dutch elm disease, or undesirable birds such as starlings and English sparrows are natives of other countries. In their native habitats these pests were less abundant and not particularly troublesome because they were kept under control by parasites or by animals preying upon them. When they are brought to a new environment favorable to their growth without their natural enemies they increase to epidemic proportions.

Australia has been particularly subject to uncontrolled population explosions of introduced species. A few prickly-pear cactus plants taken to Australia in 1840 promptly began spreading and eventually covered much of the continent because the species was not controlled by any of its natural enemies. Large areas have now been freed of the cactus by introduction of a moth that lays its eggs in the cactus, the larvae consuming and destroying the plants. Watercress introduced into Australia from England has cluttered many of the streams, even the larger ones, and made them unsuitable for navigation, while in England it is restricted to the smaller streams. Rabbits taken to Australia increased so rapidly in population that they spread over the country and destroyed many crop plants. Even widespread hunting could not control them. Introduction of a rabbit disease for a while appeared to provide good biological control, but a few individuals were immune to the disease and these are now repopulating the country.

Elimination of a species, or a great reduction in its population, can also upset a biological balance. Extensive killing of native carnivorous animals such as wolves and foxes results in a marked increase in herbivorous animals such as rabbits, squirrels, field mice, and deer. In turn these then reduce the population of the plants on which they live, completely upsetting the biological balance of the community. Furthermore, the large number of herbivorous animals then invade farms and destroy crop plants. The killing of birds and snakes that eat insects results in a great increase in the numbers of insects attacking plants, and so reduces the plant population of a community. The tiger population of Sumatra was reduced to such a degree that the wild pigs they used as food became very abundant and destroyed much of the native vegetation, including most of the young palm trees. Since the natives used palms in many ways the entire economy was disrupted.

The widespread use of insecticides can cause serious disturbances in a community. Birds and other animals that naturally control the insects may be killed by eating poisoned insects, and so still more insecticides are needed for control, with the danger of killing still more birds. The insecticides may also kill useful insects, such as the ladybird beetles that keep aphids under control, or bees that pollinate plants.

Man still has much to learn about the possibly disastrous effects of interfering with the natural system of checks and balances. However, even if we only applied what we already know the situation could be much better than it is.

Chapter 11

PLANT COMMUNITIES

We have considered the interactions of organisms in biological communities and have noted that, in a stable or climax community, the proportions of the various organisms remain constant over long periods of time. Each kind of community has a particular composition and structure. There are many different types of biological communities on the earth, each with its own characteristic species of plants and animals. On land there are forests, grasslands, deserts, and tundra. Streams, ponds, lakes, bogs, beaches, and salt marshes all have their typical communities, and there is also a variety of marine communities in the oceans. In land and most fresh-water communities the plants are the most conspicuous members and the communities are generally named for the dominant plants of the community. The dominant plants are generally the larger and more abundant ones, and are particularly characteristic of the community. In marine communities the plants are generally less conspicuous than the animals, since so many of the plants that provide the base of a marine food pyramid are microscopic, although there may be large brown or red algae present.

Plant communities can be classified into categories at various levels, just as species are. The largest and most comprehensive category is the **formation.** Among the formations found in the United States are the Deciduous Forest Formation, the Northern Evergreen Forest Formation, the Prairie Formation, and the Southwestern Desert Formation. A little later in this chapter we shall consider each of the principal formations of this country in some detail.

Each formation is composed of several rather distinct **associations.** For example, the Deciduous Forest Formation includes oak-hickory and beech-maple associations, each occurring in specific areas of the formation. Associations, in turn, are composed of a series of increasingly smaller and more uniform organizational units, but we shall not consider these in the present discussion.

FACTORS DETERMINING OCCURRENCE OF PLANT FORMATIONS

It is not just a matter of chance that in some places the natural vegetation is forests, in another grasslands, and still another deserts. In any particular region the factors of the physical environment are such that only one type of formation can exist naturally. Perhaps the most important factor in determining what particular plant formation will occupy a certain region is water availability. What counts is not just the amount of rainfall in the region, but the amount of rainfall in relation to the amount of evaporation (if water were always available to evaporate). This is expressed as the **rainfall/ evaporation ratio** (R/E ratio) calculated as $\frac{R}{E} \times 100$. If the ratio is 100 the amount of rainfall during a year is just equal to the amount that would evaporate in a year.

If two regions have the same rainfall but one is warmer than the other, the warmer region will have a lower R/E ratio because of the higher rate of evaporation. If the R/E ratio is 100 or more, forest formations will be present; if between 20 and 100, grasslands; and if less than 20, deserts. Temperature is more important than water availability in determining the extent of the tundra, however. While temperature does not play a primary role in the distribution of forests, grasslands, and desert, it is important in determining the species that will make up the formation. Thus, many of the trees in a tropical forest could not survive freezing weather in a temperate zone forest, while temperate zone trees might not be able to break dormancy or bloom without low temperature preconditioning in a tropical region even if they could stand the high temperatures.

Length of day may also influence the kinds of species present in a formation. Long-day plants cannot reproduce at or near the equator since the daylight period is never long enough, while short-day plants cannot reproduce in far northern or southern regions since it is too cold for active growth by the time the days become short enough for the plants to bloom.

Soil factors and topography generally have little influence on the kind of plant formation present in a region, but they are often important in determining what association of the formation will occupy a certain site. However, topography does have an important influence on formations when we are dealing with mountains, for the changes in temperature, R/E ratio, and other factors as one ascends a mountain are marked. It is well known that by going up a mountain one will pass through different formations in the same sequence one would if going toward the poles, except that on a mountainside each formation occupies a much narrower band.

One factor that may play a role in determining the distribution of plant communities is fire, which may occur naturally from lightning as well as through the agency of man. In particular, some regions occupied by prairie grasslands would probably be occupied by forests if there were not periodic fires that destroy young trees invading the area. While the dead and dry tops of the grasses are burned the roots are not injured and the grasses are not destroyed by the fire.

PLANT FORMATIONS OF NORTH AMERICA

The land area of North America is occupied by nine principal plant formations, and we shall discuss each of these briefly (Fig. 73). In addition, there are various aquatic formations in streams, ponds, and lakes, and several different seacoast formations, as well as various marine formations in the neighboring oceans, but we shall not consider these here. In general, the native vegetation of eastern North America originally consisted of several forest formations, while the area between the Mississippi River and the Rocky Mountains was largely occupied by grasslands and, in the southwest, by deserts. In the

mountains and west of them were other forest formations, with deserts to the south. In the far north the tundra remains. The past tense has been used deliberately, since man has destroyed vast areas of the native vegetation by clearing land for cultivation, cities, and highways. However, there are still at least remnants of each of the natural plant formations.

The Tundra. The northern part of Canada and Alaska and the unglaciated edges of Greenland are occupied by the tundra, and there are southward extensions of the tundra in the high western mountains. The soils of the tundra are shallow and poorly drained, generally with permanent ice only a few inches below the surface. Poor soil aeration is common. The growing season is very short and most of the year it is intensely cold with violent dry winds and light snows. Light is continuous or nearly so during the growing season, while there is continuous or almost continuous darkness during the winter. The vegetation of the tundra is composed of low-growing xerophytic plants with shallow roots, leathery leaves, and creeping stems. There are various grasses, sedges, mosses, lichens, and low shrubs including cranberries, crowberries, snowberries, and dwarf willows. The animal members of the Tundra Formation include the Arctic fox, Arctic hare, musk ox, caribou, bears, bighorn sheep, goats, ducks, geese, and numerous flies and mosquitoes. During the summers the mosquitoes are so abundant that they make life quite miserable for anyone who has a reason for being in the tundra.

The Northern Evergreen Forest. All of eastern Canada south of the tundra, the northern part of the New England states, the central region of western Canada, and substantial parts of Alaska are occupied by the Northern Evergreen Forest. The dominant trees are spruces, pines, and hemlocks together with considerable birch and aspen, particularly in cut-over areas. The region just to the north of the Great Lakes is occupied principally by a pine-hemlock association which is sometimes set aside as a separate formation (the Lake Forest Formation) but is more commonly considered to be part of the Northern Evergreen Forest Formation. The Northern Evergreen Forest extends southward into the Carolinas in the higher parts of the Appalachian Mountains. It formerly extended farther south than it does now, and the trees can grow well farther south than they do, but are excluded by the trees of the Eastern Deciduous Forest. The Northern

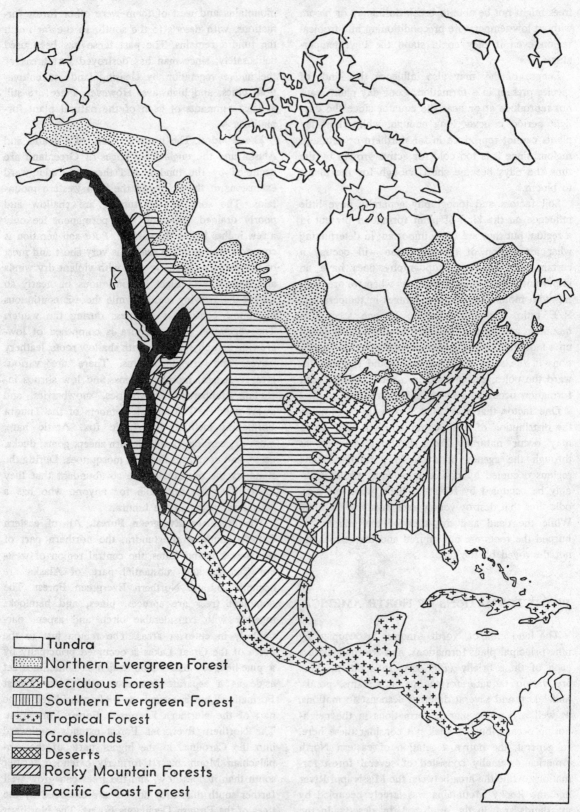

Tundra
Northern Evergreen Forest
Deciduous Forest
Southern Evergreen Forest
Tropical Forest
Grasslands
Deserts
Rocky Mountain Forests
Pacific Coast Forest

Fig. 73. The plant formations of North America.

Evergreen Forest is gradually extending northward into areas previously occupied by tundra, probably as a result of a long-range and gradual increase in temperature.

The region is characterized by long and cold winters and a short but warm growing season of three to four months. The humidity is high and there are deep snows and considerable rain, with a R/E ratio of over 100. Because the area was all glaciated the soils are shallow with rock not far below the surface and so the drainage is rather poor. Among the animals of this formation are deer, moose, woodland caribou, elk, wolves, black bears, and red squirrels.

Eastern Deciduous Forest. The Eastern Deciduous Forest (sometimes called simply Deciduous Forest) formerly occupied practically all the area south of the Northern Evergreen Forest and east of a line extending south from central Minnesota and Iowa, along the western border of Missouri, and through east Texas, *i.e.*, at about 95° longitude. Excepted were the southern third of Florida, the southeastern coastal plain that is occupied by evergreens, and a few patches of prairie extending through Illinois and into Indiana and even Ohio. The principal associations are beech-maple, oak-hickory, and oak-chestnut, though the chestnut trees have almost all been killed by the chestnut blight. Other common species are elms, ashes, sycamore, cottonwood, and tulip poplar. The upland areas with poorer soils are occupied by the oak-hickory formation, while the richer upland soils have the beech-maple formation. In the lower and more poorly drained areas and in valleys elm, ash, soft maple, birch, and sycamore are common. A wide variety of herbaceous plants, principally spring wild flowers, and of shrubs grow under the trees. Of the smaller trees flowering dogwood is particularly conspicuous and abundant.

The region is moist, with rainfall/evaporation ratios between 100 and 110, and the soils are generally quite deep and fertile. The growing season is long and the winters rather short, but cold. Among the original animals of the formation were Virginia deer, bobcat, black bear, wild turkey, mink, muskrat, foxes, raccoon, opossum, squirrels, skunks, rabbits, and a variety of birds and reptiles.

Southern Evergreen Forest. The rather broad coastal plain from Virginia to Texas is occupied by the Southern (or Southeastern) Evergreen Forest. On the sandy uplands the principal trees are pines, especially longleaf and shortleaf pines. The swamps

are frequently occupied by cypress trees with the characteristic cone-shaped knees extending upward from their roots. Along the banks of streams and bayous are sour gum, magnolias, water oaks, and live oaks, with Spanish moss frequently draped from their branches. On the low hills of the Mississippi flood plains there are extensive canebrakes. On hummocks, trees of the deciduous forest such as oaks, hickories, and beeches invade the region. There are at least one or two killing frosts each year, but the winters are short and mild and the summers long and hot. The soils are generally sandy and not very fertile. The region is very moist, with R/E ratios usually above 110. In addition to many of the same animals that live in the Deciduous Forest there are numerous reptiles such as snakes and alligators, and birds like egrets that are not found in the Deciduous Forest.

Tropical Forest. The southern third of Florida, the West Indies, the very tip of Texas, and the lowlands of Mexico and Central America are occupied by the Tropical Forest. Among the principal trees are palms and mangroves, as well as a variety of tropical species whose names would mean little to those not acquainted with them. Rather extensive areas are also covered with palmettos, a dwarf species of palm. The Florida Everglades are in this region. Some regions are occupied by jungle, but this is secondary growth following clearing rather than an original association. Many epiphytes, including different orchids and ferns, grow on the trees. The region is characterized by high humidity and high temperatures. Freezing weather is rare. The characteristic animals include opossums, shrews, bats, jaguars, guinea pigs, porcupines, and anteaters.

Prairie Grasslands. The Prairie Grasslands (or simply Grasslands) occupy a long narrow strip, principally between longitudes 95° and 100° or from about the middle of Iowa to the middle of Nebraska. Northward the prairies extend into central Canada and southward to the Texas coast. Patches of prairie are abundant in Illinois, rarer in Indiana, and very rare in Ohio. These prairie patches are remnants of a prairie peninsula that extended into this region about three thousand years ago when the climate was much drier than at any other time. In the eastern part of the Prairie Grasslands, forests are present along the streams and on some of the better-drained areas. Fires are common in the prairies in the fall when the tall grasses are dead and dry, and this is one factor that has kept the trees from extending into the eastern prairies. Since most of the prairie is now under cultivation and since fires are now better controlled, trees are growing in regions once occupied by grasslands. In the western part of the prairies it is too dry for successful invasion of the trees.

The dominant plants of the Prairie Grasslands are various species of tall grasses, three to ten feet tall, but there are also numerous herbaceous plants with bright and conspicuous flowers, particularly in summer. The soil is deep and fertile with much humus, but is frequently poorly drained. Periods of heavy rainfall and severe droughts alternate, but in general the R/E ratio is low—between 60 and 80. The region is not flat as many people who have not seen it suppose, but rather quite hilly. Among the native animals are bison, pronghorn sheep, wolves, coyotes, jack rabbits, prairie dogs, prairie chickens, and grouse.

Plains Grasslands. The region west of the prairies and east of the Rocky Mountains and from central Canada through Texas is occupied by the Plains Grasslands. This formation also occupies some areas west of the Rockies. In the southwest it grades into desert. In contrast with the prairies the dominant grasses are short, rarely over a foot high, and include such species as buffalo grass, bunch grass, and wire grass. There are also many plants with conspicuous flowers, principally legumes, composites, and yucca. In the southwest, cactus and sagebrush are present; sagebrush is also found in Colorado and Montana. The soils are rather heavy, and may be somewhat alkaline. Rain is quite irregular and the total annual rainfall is low, the R/E ratio being 20 to 60. In the late summer there are hot, dry winds from the southwest. It is in the plains region that dust storms are so common. Much of the region is under cultivation and most of the rest is grazed. It seems likely that the entire plains region would best have been used as grazing land since this would have prevented most of the dust storms. Besides, in many years it is too dry for successful crops. However, in moist years the yields are high as the soils are generally quite fertile. The animals of the Plains Grasslands are in general rather similar to those of the Prairie Grasslands.

Southwestern Desert. The Southwestern Desert is found in southern California, Nevada, Utah, Arizona, New Mexico, and southern Texas. It also extends into Mexico both east and west of the mountains and occupied essentially all of Lower California. The various species of cactus are commonly thought of as the main desert plants, and they are abundant. But there are also many other species including agaves, yuccas, Joshua trees, mesquite, a variety of spiny or thorny shrubs, and many small annuals with bright and conspicuous flowers that complete their entire life cycle during the brief rainy seasons. To the north the sagebrush deserts predominate and grade into the Plains Grasslands. Deserts are sometimes thought of as being barren of plants, and some like the Sahara are essentially so except for oases. Our southwestern deserts, however, are well populated with a great variety of bizarre plants.

Deserts occur where the rainfall/evaporation ratio is under 20, a result both of low rainfall and a high rate of evaporation. During the summer the temperature commonly is high, frequently up to 120° F., but there may be frosts in the winter and the northern parts of the desert may get quite cold. The rains come principally during two times of the year—January and July. The soils are generally sandy and quite alkaline, with a high mineral salt content. The desert animals are quite characteristic and include many reptiles such as various snakes, lizards, horned toads, and gila monsters. There are also numerous rodents, wolves, and coyotes. Large animals are rare but there are several species of deer.

Western Evergreen Forest. The Western Evergreen Forests are found in the mountains and also along the West Coast. There are a number of quite distinct associations that are sometimes grouped into several formations, rather than the one as we have done. The evergreen forests extend from Canada into Mexico. Along the coast the dominant species from Alaska, through Canada, and into Washington is the Sitka spruce, but in Washington and Oregon, Douglas fir is dominant. These are among the tallest of all trees, reaching heights of as much as 400 feet. Large arborvitae, cedars, and firs are also present. In southwestern Oregon and northern California, redwoods or sequoias are the dominant trees. These huge trees grow to 340 feet in height and 20 feet in diameter. Under them is a luxuriant undergrowth of ferns, herbaceous plants, and shrubs, including rhododendron and dogwood. These coastal forests are not only the tallest, but also the most luxuriant in the country. The region is characterized by heavy rainfall and a very high R/E ratio, commonly over 120. The soils are deep and fertile and the climate is moderate even as far north as Washington as a result of warm ocean currents, the winters being quite warm and the summers cool and humid.

The southern California coast is occupied by quite a different type of evergreen forest called chaparral. This consists of rather small evergreen oaks with hard and leathery leaves and other species with similar xerophytic structure. In contrast with the northern coastal forests the R/E ratio is quite low and there are definite dry and rainy seasons. Around Monterey Bay the cypress trees, often quite distorted by wind and salt spray, are the characteristic species.

In the mountains the principal trees are western yellow pine, western white pine, and Douglas fir. Other common conifers are western larch, western hemlock, white fir, red cedar, sugar pine, lodgepole pine, and Engelmann spruce. Farther south, particularly in the Great Basin between the Rockies and the Sierras and on the Arizona plateau, the piñon-juniper woodlands association replaces the characteristic mountain associations. It consists primarily of fairly open stands of two species of piñon pine and several species of juniper, all of them relatively small trees. To the south it grades into scrub oak chaparral and then into grasslands and semidesert.

The animals vary from one plant association to another, but in general the area includes grizzly bears, black bears, mule deer, bighorn sheep, mountain lions, mountain goats, and timber wolves.

PLANT SUCCESSION

We have described the various plant formations on the basis of their **climax associations,** *i.e.,* associations that remain unchanged in structure and composition as long as they are undisturbed and there are no major environmental shifts. We know, however, that the plant associations occupying a certain area have not always been there. Forests once occupied the areas where the southwestern deserts now exist; prairies once occupied Illinois, Indiana, and Ohio; palm trees once flourished in Greenland; many areas now dry land were once at the bottom of the ocean and vice versa. Such marked changes in the vegetation of an area result from major climatic and geological changes. Considered from a long-range geological standpoint, plant evolu-

tion also has played a role in marked changes in the vegetation. For example, during the coal ages the forests consisted of huge tree ferns, club mosses, and horsetails—species that are now extinct—rather than our present trees that had not yet evolved.

Such replacement of one plant formation by another, which probably occurred gradually and in successive stages rather than suddenly, could be considered as **plant succession.** However, the term is usually used for somewhat less marked and dramatic changes in the associations occupying an area within a formation that occur in hundreds (or perhaps a few thousands) of years rather than over longer periods of time. This more usual type of plant succession occurs when there is a bare area that can be invaded by plants. Natural or primary successions may begin on rock outcrops or on land formed by the silting in of ponds or lakes, while secondary successions occur when land that man has cleared is allowed to revert to natural vegetation.

In the Eastern Deciduous Forest region succession on a bare rock surface generally begins with lichens, mosses, and crevice herbs, passes through perennial herb, shrub, and pine stages to an oak-hickory forest, and then finally to a climax beech-maple forest (although the oak-hickory forest may be the climax association in some areas). Such a succession is made possible by the gradual increase in the soil as the rocks are eroded and weathered and as each successive community contributes humus to the soil. The pine forest also provides shade and favorable water relations that permit growth of the hardwood seedlings, while at the same time the seedlings of the pines themselves cannot survive in the relatively low intensity light of the pine forest floor.

As a pond or small lake begins filling in, bulrushes and cattails become established and contribute to further filling in and holding of the soil, thus providing conditions favorable for a sedge swamp. As soil building continues, alders, willows, wild roses, and other plants invade the area, to be followed by an elm-ash-soft maple forest. Finally, when the soil-building process is complete and the soil has attained good enough aeration the climax beech-maple forest becomes established. Each succeeding association changes the environment in ways that make it more suitable for the next association in the series and less suitable for itself. Thus, the climax beech-maple association may be the final result of two successions starting from quite different habitats—

a bare rock surface and a pond or lake. In the first case the succession leads to a moister and moister environment as soil accumulates, and in the second to a drier and drier environment. Both result in a habitat with intermediate moisture and adequate soils, favorable to the beech-maple forest.

More rapid and easier to follow than either of these natural successions are secondary successions that occur in abandoned fields. The old field successions that occurs in the Piedmont Region of North Carolina and adjacent states has been thoroughly studied and described in some detail. The entire succession is completed within a period of two hundred years or less. Crab grass is generally the first plant to invade the field, and may appear even while the crop is being raised. It survives through the first two years after abandonment of the field, but then dies out because of increasing shade. The first year the field is abandoned it is thickly covered with horseweed, but the second year wild aster takes over and shades out the horseweed, since the environment was then favorable for the growth of the aster seedlings. The third year broom sedge, in turn, replaces the asters and continues to be present for many years, though from the fourth year on many shrubs appear and pine seedlings begin growing. By twenty years after abandonment of the field a young pine forest has become established and generally becomes so dense that it shades out most of the grasses and shrubs that require bright light. The young pine seedlings are also unable to survive in the rather dim light of the pine forest and at first there is little undergrowth.

The seedlings of oaks, hickories, and other hardwoods are able to grow under the pine trees whereas they cannot survive in the open, and some 30 years after the field was abandoned young hardwood seedlings begin being evident. However, until

75 years or so the forest is definitely pine with a hardwood understory. Since many fields were abandoned anywhere from 25 to 100 years ago there are many pine forests in the Piedmont Region and some people mistakenly assume that this forest is part of the Southeastern Evergreen Forest. By 100 years the pines have begun to die out and the oaks, hickories, and other hardwoods have become large enough that the forest may be considered a mixed pine-hardwood forest. Few if any pines remain after 125 years, and by 150 years they are all gone and the climax oak-hickory forest has become established. Some beeches and hard maples may come in, but in this part of the Eastern Deciduous Forest the climax is oak-hickory rather than beech-maple. The forest includes a variety of other hardwoods in addition to oaks and hickories, and underneath the trees are a good many dogwood trees and a variety of shrubs and wildflowers. There are also seedlings of the hardwood trees that will keep replacing the larger trees as they die and so maintain the climax association. The Piedmont Region contains many examples of each of the various stages in the succession, and if one knows the sequence it is possible to estimate quite accurately how long the land has been out of cultivation. There are few virgin forests in the region, since almost all the land has been in cultivation at one time or another and so virtually all the oak-hickory forests are the products of secondary succession.

The pine trees are more valuable commercially than the hardwoods, and so owners of woods in many cases attempt to prevent the succession to the climax oak-hickory forest. One way that this can be done is by selective cutting of the pine trees. This admits enough light to the forest floor so that the pine seedlings can survive and the hardwood seedlings cannot, and so maintains the pine forest.

Chapter 12

PLANT REPRODUCTION

The capacity of each species of living organism for reproducing its kind is one of the best and most universal distinctions between living and nonliving things, and it is also a most remarkable process when one stops to think about it. Since all living organisms eventually die, reproduction is essential for maintenance of the species. Reproduction also provides the means of increasing the population of a species and the means of hereditary variation. Evolution of new species could never have occurred without reproduction. Since most plants are immobile, reproduction also plays another important role in the plant kingdom—the dispersal of a species from one place to another, generally by means of spores or seeds.

Reproduction is basically a matter of the detachment of one or more cells from the parent and their growth into a new individual. In **asexual reproduction** the detached cells grow into new individuals directly, but in **sexual reproduction** the fusion of two cells into one is essential before growth can proceed. In most plants and animals the fusing cells are an **egg** and a **sperm,** generally produced by a female and a male individual respectively. However, in some of the simpler organisms the two cells that fuse are identical and may be produced by a single individual, or even if eggs and sperm are produced they may both come from a single individual. Thus, sexual reproduction may exist without male and female individuals or even without distinct eggs and sperm, but not without fusing cells. In general, the cells that fuse in sexual reproduction are called **gametes,** whether they are distinct eggs and sperm or not, and the cell resulting from the fusion of two gametes is called a **zygote.**

ASEXUAL REPRODUCTION

The vegetative propagation of vascular plants by means of stems, roots, or leaves, discussed in Chapter 3, represents one general type of asexual reproduction in plants, whether it occurs naturally or through the agency of man. Here large groups of cells—tissues or even whole organs—are detached from the parent and develop into a new individual.

Among the algae and fungi there are various means of asexual reproduction. Some of these involve the detachment of groups of cells from the parent, and so are essentially comparable with the vegetative propagation of vascular plants. However, several other specific means of asexual reproduction are more characteristic. Unicellular bacteria, algae, and fungi commonly reproduce by **fission** (Fig. 74).

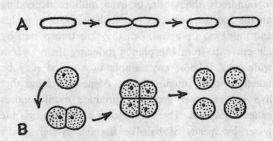

Fig. 74. Asexual reproduction by fission. **A**–A bacterium. **B**–A unicellular alga.

This is simply cell division, the cells becoming detached from one another after division and so constituting two new individuals. Thus, in unicellular organisms, cell division is a means of reproduction, rather than growth of the individual. The parent essentially disappears in the process of reproduction, so there is no old age or natural death, but at the same time the offspring are essentially orphans.

Yeasts commonly reproduce by a modified form of fission known as **budding** (Fig. 75). The new cell

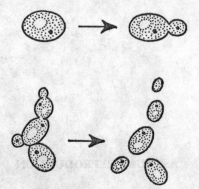

Fig. 75. Asexual reproduction of yeast by budding.

is smaller than the parent cell, and so the parent retains its identity. The smaller bud cell may in turn divide while still attached to the parent cell, and so there may be a short chain of cells of decreasing size. Eventually, however, the cells become detached and each is then an individual yeast plant.

The most universal type of asexual reproduction among the algae and fungi is sporulation (Fig. 76). Spores are single cells (occasionally groups of two to a few cells) that become detached from the parent and under favorable conditions grow into new plants. Each individual generally produces a large number of spores, ranging from a few dozen to hundreds, thousands, or even millions depending on the species. Some spores, such as those of molds and mildews, are aerial, *i.e.*, they are transported by air currents from the plant producing them over a wide area. Almost any sample of air that may be taken contains fungus spores. Algae and fungi that live in the water generally produce motile spores called **zoospores**. These zoospores swim through the water by means of hairlike flagella. Aerial spores are generally produced in much larger numbers than zoospores.

Unfortunately the term "spore" has been used for a variety of other things besides the true asexual spores just discussed, including the hard-walled resting spores of bacteria (which play a role in survival through unfavorable periods rather than in repro-

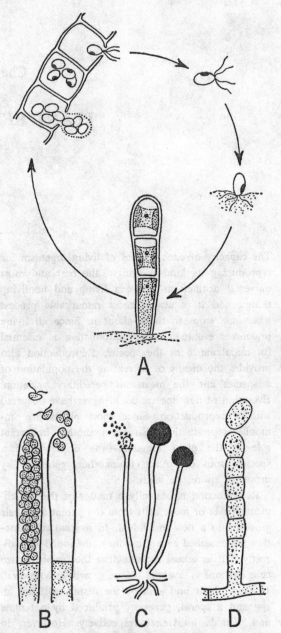

Fig. 76. Asexual reproduction of some algae and fungi by sporulation. A–Asexual reproduction of the alga *Ulothrix* by zoospores. B–Sporangia and zoospores of the water fungus *Saprolegnia*. C–Black bread mold with sporangia, one of which has opened releasing aerial spores. D–Production of aerial spores (conidia) by a mildew.

duction), the zygospores which are produced by the fusion of the sex cells of some fungi and algae, and meiospores, which play an essential role in the sexual reproduction of plants and will be discussed later. Some of the spores of fungi and algae, as well as all of the spores produced by the bryophytes and vascular plants, are meiospores. When we use the term "spore" here with no further identification it applies to the true asexual spores.

SEXUAL REPRODUCTION

Although many species of plants have some means of asexual reproduction, most species of plants can reproduce sexually. The bacteria and blue-green algae are the only major groups of plants without typical sexual reproduction in at least most species, and even the bacteria (and perhaps the blue-green algae) have means of interchanging genetic material between individuals that some biologists regard as a very simple type of sexual reproduction. A few higher plants such as bananas are sterile and cannot produce seeds, even though they have flowers, and so are dependent on asexual reproduction by vegetative propagation.

In any sexual life cycle there are two important and essential events: one is the union of the gametes, or **fertilization,** and the other is **reduction division,** a particular type of cell division in which the number of chromosomes in the parent cell is reduced by half in each of the offspring cells. Reduction division always consists of two successive divisions, resulting in four cells from the parent cell. The first division is the one in which the chromosome number is reduced by half and is known as **meiosis,** in contrast with mitosis, the usual type of cell division (Chapter 8) in which the offspring cells have just the same numbers and kinds of chromosomes as the parent cells. The second division, occurring in each of the two cells produced by meiosis, is a mitotic division.

When a sperm fertilizes an egg, each contributes one complete set of chromosomes to the zygote, which then has two complete sets. The number of chromosomes per set is characteristic of a species, and may range from only one or two to one hundred or so, though the number is most frequently from eight to about twenty. Gametes and other cells with one set of chromosomes are said to have n or **haploid** number, while zygotes and other cells with two sets have $2n$ or **diploid** number of chromosomes. All cells produced by mitotic divisions of the zygote, as in ordinary growth of the zygote into an individual, have the $2n$ or diploid number of chromo-

somes. However, before gametes are produced reduction division occurs. If it did not, the gametes would contain the $2n$ number of chromosomes and the zygote produced by their fusion would contain the $4n$ number. Without reduction division, the number of chromosomes would double each generation. Thus, both reduction division and fertilization are essential in sexual reproduction if the number of chromosomes is to be maintained at a constant number generation after generation, as it ordinarily is.

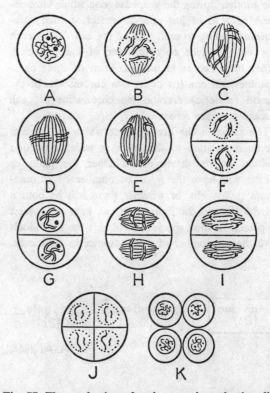

Fig. 77. The production of meiospores by reduction division. A–Spore mother cell in interphase. B–Prophase of first division, showing the four chromosomes that constitute the $2n$ number. Each chromosome has duplicated itself and consists of two joined chromatids. C–The homologous chromosomes synapse with each other. D–The synapsed chromosomes line up at the equator at the metaphase. E–During anaphase the two homologous chromosomes of each pair separate, one moving to each pole. F–Early telophase, each of the resulting cells containing only two chromosomes (one set), the n number. G–Interphase. H–Metaphase of second division. I–At anaphase the chromatids of each chromosome separate, each of the four resulting meiospores now having two chromosomes (J). K–Mitosis and cytokinesis are complete, the result being four separate meiospores.

The process of meiosis as it occurs in reduction division is diagrammed in Fig. 77. A complete description of the process is not necessary here since most stages are the same as in mitosis. However, attention should be called to the following essential differences that result in reduction of the chromosome number by half, in contrast with its maintenance at the number in the parent cell by mitosis: 1. in meiosis the comparable (homologous) chromosomes of the two sets pair off with one another during the prophase, a process known as **synapsis**; 2. the paired or synapsed chromosomes line up at the equator during the metaphase and separate from one another during the anaphase, one whole chromosome (instead of just one chromatid of each chromosome) going to each end of the cell. Thus, each new cell receives only one set of chromosomes, rather than two as in mitosis. Each of these chromosomes still consists of its two chromatids. In the second (meiotic) division the chromatids of each chromosome are separated.

The point in the life cycle at which fertilization occurs in relation to the point at which reduction division occurs determines whether the plant or animal will have the n or $2n$ number of chromosomes in its cells, or whether there will be both n individuals and $2n$ individuals, as in most species of plants. In most species of animals the individuals have the $2n$ number of chromosomes in their cells.

Reduction division occurs in the formation of the eggs and sperm, so only the gametes have the n chromosome number. As soon as fertilization occurs the number goes back up to $2n$ in the zygote and all cells derived from it by mitosis. In some species of algae and fungi reduction division occurs just after fertilization, rather than just before fertilization as in animals. That is, the first two divisions of the zygote are reduction divisions. The result is that the cells of the plants that grow from the four cells produced all contain the n number of chromosomes. In most species of plants, however, fertilization and reduction division are quite widely spaced. The $2n$ zygote divides by mitosis, as in animals, and so produces a $2n$ individual. When this individual is mature it carries on reduction division. However, the four cells produced from each reproductive cell are not gametes (as in animals) but rather meiospores (*i.e.*, spores produced by meiosis). The meiospores then divide by mitosis, each one giving rise to an individual with the n number of chromosomes in its cells. When this n or haploid individual is mature it produces gametes. Thus, in most plants a generation of n or haploid individuals reproducing sexually by means of gametes (the **gametophyte** generation) alternates with a generation of $2n$ or diploid individuals reproducing asexually by means of meiospores (the **sporophyte** generation). This alternation of generations is characteristic of the plant kingdom,

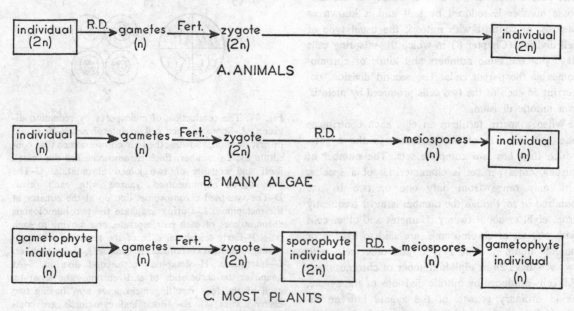

Fig. 78. Comparison of typical sexual life cycles of animals, many algae, and most plants.

being found in most species of algae and fungi as well as in higher plants, and is essentially absent from the animal kingdom. The gametophyte and sporophyte generations are generally quite different from one another in size, structure, and general appearance. Fig. 78 outlines the characteristic life cycles of animals, the haploid algae and fungi, and the alternation of generations of most plant species.

The remainder of this chapter will be devoted to a consideration of the life cycles of representative species from some of the major plant groups in an effort to clarify the concept of alternation of generations and to point out the evolutionary advances that have occurred in the plant kingdom in relation to the sporophyte and gametophyte generations. Despite the modifications and complications that occur in the various representative life cycles, you should note that all the life cycles fit into the generalized schemes outlined in Fig. 78.

REPRESENTATIVE LIFE CYCLES

The life cycles selected for consideration here include representatives of the green algae and of all the more advanced groups of green plants including the bryophytes, ferns, club mosses, gymnosperms, and angiosperms. This set of life cycles samples what may be considered as the main line of evolution of reproduction in the plant kingdom, and omits the many different phyla of algae and fungi that have evolved in various specialized directions. In many of these, particularly some of the fungi, there are very complex and involved methods of reproduction; but in general most of them have a more or less typical alternation of generations.

Ulothrix. As the first representative of the green algae we take *Ulothrix*, a filamentous fresh-water genus that is widely distributed (Fig. 79). All the

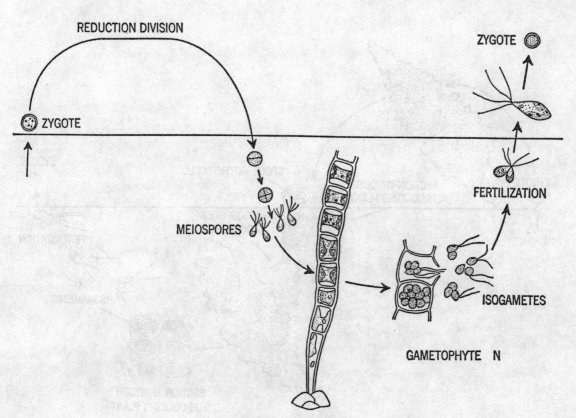

Fig. 79. Life cycle of *Ulothrix*.

cells of the filament are essentially similar except the basal one, which serves to anchor the filament to a rock or other submerged object and is known as the **holdfast.** *Ulothrix* reproduces asexually by means of zoospores, up to thirty-two of them being formed in a cell by internal mitotic divisions. Each zoospore has four flagella, and when the zoospores are mature the cell breaks open and the zoospores emerge, swim around for a while, and then settle down and grow into a new filament. The zoospores, as well as all the cells of the filament, have the *n* or haploid chromosome number.

Ulothrix also produces isogametes, up to sixty-four being formed in a cell. They resemble zoospores except that they are generally smaller and they have two flagella instead of four. After they have emerged from the filament they cannot grow into a filament directly as do the zoospores, but must first fuse with another isogamete. The product is a 2*n* or diploid zygote which initially has four flagella and swims around. Later it loses its flagella and becomes encased in a thick wall. In this state it may remain alive for considerable periods of time, even

through unfavorable environmental periods such as the drying up of the water in which the plants were living. The zygote germinates by dividing into four zoospores, which are really meiospores since they are produced by reduction division. The zoospores may then develop in new plants. *Ulothrix* plants are thus haploid gametophytes and a diploid sporophyte generation is lacking except for the zygote. Although *Ulothrix* has sexual reproduction it has neither separate sexes nor differentiated eggs and sperm. However, gametes generally fuse only with those from another plant so there may be some sexual distinction.

Oedogonium. The life cycle of *Oedogonium,* another filamentous, fresh-water green alga, is similar to that of *Ulothrix* except that it is **heterogamous** rather than **isogamous,** *i.e.,* it has distinct eggs and sperm. This is a definite evolutionary advance. In some species both eggs and sperm are produced by one individual, but in others there are male plants that produce only sperm and female plants that produce only eggs, a further evolutionary advance. As in *Ulothrix,* the zygote is the only diploid stage,

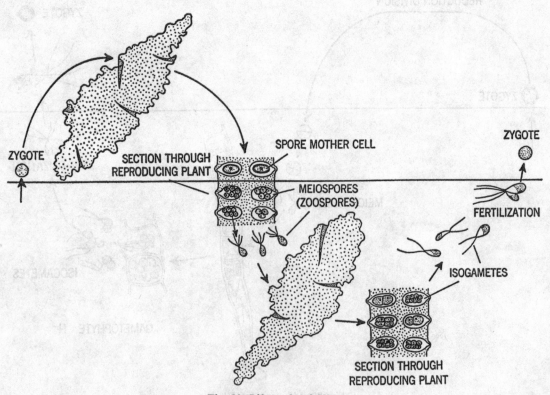

Fig. 80. Life cycle of *Ulva.*

reduction division occurring when it germinates into zoospores. The evolution of the large non-motile egg and small motile sperm makes possible a zygote with a large supply of accumulated food without the considerable expenditure of energy that would be required in the swimming of a large motile gamete.

Ulva. The life cycle of *Ulva* (sea lettuce) is included as a representative of the green algae that have a true alternation of generations with definite gametophyte and sporophyte generations (Fig. 80). Sea lettuce is a common marine alga that grows in the tidal zone of the seacoast. Both the gametophytes and sporophytes are leaflike sheets two cells thick and several inches to a foot or more in length with holdfasts that attach them to rocks or stones. Isogametes produced within the cells of the haploid gametophytes emerge when mature, swim around for a while, and then fuse in pairs producing xygotes. These zygotes, instead of undergoing reduction division immediately as in *Ulothrix* and *Oedogonium,* grow into a filament of diploid cells. The filament then develops into the sheetlike mature sporophyte.

Cells along its margin serve as sporangia, the first division being meiotic and the remaining divisions that result in four or eight zoospores (meiospores) per sporangium being mitotic. Thus, reduction division results in the production of haploid zoospores by the diploid sporophyte. The zoospores grow into filaments which then develop into the sheetlike mature gametophyte. The formation of the filaments in the development of both the gametophyte and sporophyte plants suggests that the ancestors of sea lettuce were filamentous algae and is an example of **recapitulation,** the appearance in development of stages similar to those the species passed through in its evolution. Although isogamous, gametes fuse only with those from another plant.

Moss. The green, "leafy" moss plants that make up the cushionlike mats in which moss usually grows are gametophytes and have the haploid chromosome number in their cells. At the tips of the leafy stalks are clusters of multicellular structures within which the gametes are produced, a single egg in each flask-shaped **archegonium** and many sperm within each club-shaped **antheridium** (Fig.

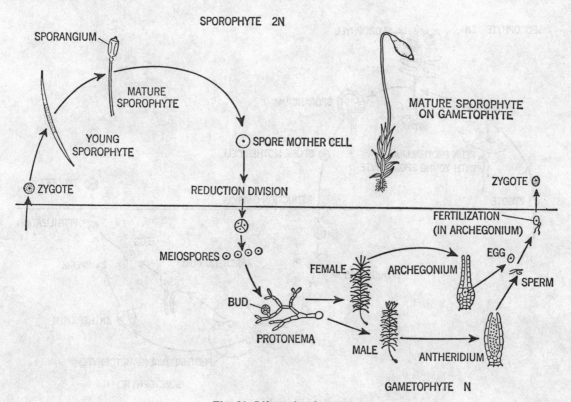

Fig. 81. Life cycle of a moss.

81). If terms applied to animals were used, the archegonia would be called ovaries and the antheridia testes. Antheridia and archegonia may be borne together in a mixed cluster, on separate branches, or on separate individuals, depending on the species. Only in the latter case are these separate male and female plants. The flagellated sperm swim out of the antheridium, through the film of water that frequently covers mosses, and through the neck of the archegonium to the egg. Fertilization then occurs and the diploid zygote begins developing into the sporophyte plant.

The sporophyte consists of a hairlike stalk bearing a **sporangium** (capsule) at its upper end. Within the capsule certain cells become distinguishable as spore mother cells. Each of these undergoes reduction division, producing four meiospores. When the spores are mature the lid of the sporangium opens and the spores are dispersed by air currents, sometimes to considerable distances. If they land in a suitable environment they germinate into a branched filamentous young gametophyte plant (another example of recapitulation, indicating algal ancestry). Buds on

the filament later grow into the characteristic leafy, upright branches.

In mosses the gametophyte generation is the principal one (as in most algae), carries on most of the photosynthesis, and is continuously present in contrast with the seasonal sporophytes. As far as reproduction is concerned, the most important evolutionary advance of mosses (and liverworts) over the green algae is that mosses have multicellular reproductive organs (sporangia, antheridia, and archegonia) in contrast with the unicellular gametangia and sporangia of the algae. In both mosses and liverworts, and also in the vascular plants, all species are **heterogamous** (*i.e.*, produce eggs and sperm), all isogamete production being restricted to some species of algae and fungi.

Fern. In contrast with the mosses and liverworts, the conspicuous generation of the ferns (and all other vascular plants) is the sporophyte rather than the gametophyte (Fig. 82). In vascular plants the sporophyte has roots, stems, and leaves containing vascular tissue (xylem and phloem) and may attain great size. The spores of most of the common

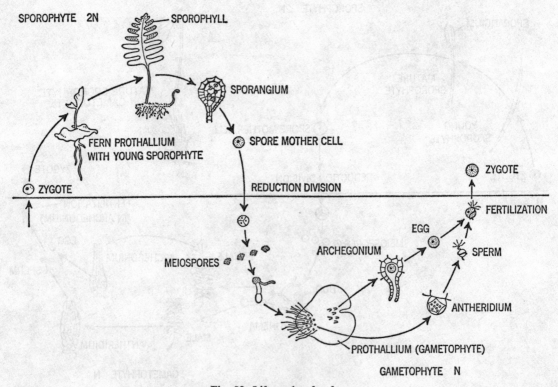

Fig. 82. Life cycle of a fern.

species of ferns are borne on the undersides of the leaves, the sporangia being borne in clusters known as **sori.** The leaves are known as **sporophylls,** since they bear spores. Some species of ferns have sporophylls quite different from the ordinary foliage leaves. These are generally covered with sori and usually lack extensive blades and chlorophyll.

The cells within the walls of the sporangia become spore mother cells that undergo reduction division, each mother cell producing four haploid meiospores. At maturity the sporangia snap open and the spores are shot into the air. They may be carried some distance by the air currents before landing. If the environment is favorable they germinate and grow into a small, heart-shaped gametophyte plant that is generally less than a quarter inch across. This leaflike gametophyte has chlorophyll and makes its own food. On its underside are hairlike outgrowths **(rhizoids)** that aid in absorption, a cluster of a few archegonia near the notch of the heart, and a cluster of a few antheridia near the point of the heart.

As in mosses, each archegonium contains a single egg. The flagellated sperm swim from the antheridia into the archegonia and fertilize the eggs. The diploid zygote that results then grows into a young sporophyte. The gametophyte dies soon after it has reproduced. Few people have seen fern gametophytes, since they are so small and inconspicuous and are generally hidden among the leaves and litter on a forest floor.

Club Moss. While the club mosses are not on the direct evolutionary line to the seed plants, they exhibit evolutionary advances over the ferns in reproduction that suggest stages leading to the evolution of seeds as an important part of the sexual life cycle (Fig. 83). The club moss sporophytes have numerous small leaves and very short internodes. In some species the ordinary foliage leaves are also the sporophylls, but more typically the sporophylls are specialized and arranged in cones at the tips of the stems (the "clubs" giving the group its name). These sporophylls soon turn yellow, in contrast with the green foliage leaves. A sporophyll

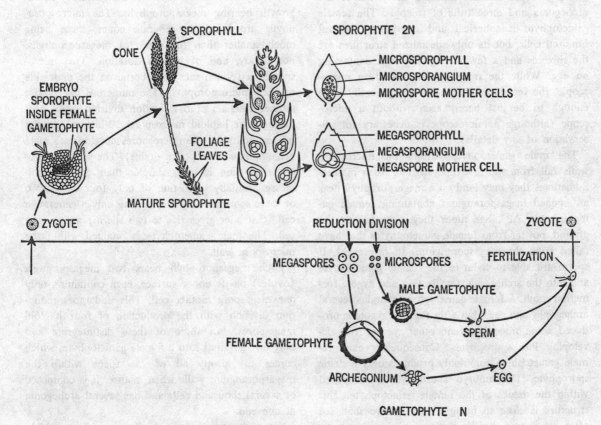

Fig. 83. Life cycle of a club moss.

generally bears only a single sporangium on its upper side. In the genus *Selaginella* there are two kinds of sporangia: the **megasporangia** and the **microsporangia.** The megasporangia produce only four large **megaspores,** which always develop into female gametophytes. The microsporangia produce many small **microspores** which always develop into male gametophytes. The sporophylls bearing the two kinds of sporangia are called **megasporophylls** and **microsporophylls.** Thus, in *Selaginella,* there is sexual differentiation in the sporophyte generation, even though it reproduces asexually by means of spores.

The male and female gametophytes differ from each other greatly in size and structure, but both are small and inconspicuous and the male gametophytes lack chlorophyll. The male gametophytes develop within the microspore wall, and consist of nothing but a single antheridium containing numerous flagellated sperm plus one vegetative cell that constitutes the whole vegetative plant body. The female gametophyte also develops within the spore wall, but the megaspore wall splits open when the female gametophytes are mature, exposing the necks of the archegonia and three tufts of rhizoids. The female gametophyte is spherical and consists of several hundred cells, but its only specialized structures are the rhizoids and a few archegonia, each containing an egg. While the male gametophytes are microscopic, the female gametophytes are generally large enough to be just barely seen without a microscope (although a microscope is necessary for observation of any details).

The male gametophytes within the microspore walls fall from the microsporangium when mature. Sometimes they may land on a megasporophyll near an opened megasporangium containing female gametophytes. At other times they may fall to the ground, not far from female gametophytes that have fallen from the megasporangium. In any event, if sperm are able to swim to the female gametophyte and into the archegonia, fertilization and zygote formation result. A female gametophyte contains several archegonia with eggs, but after the first zygote is produced, some process prevents other zygotes from developing into sporophytes. Consequently each female gametophyte commonly produces only a single sporophyte. The embryo sporophyte is imbedded within the tissues of the female gametophyte. This structure is close to being a seed, as we shall see after we discuss the life cycle of a pine. However,

the embryo sporophyte cannot remain alive long without growing into a mature sporophyte of the club moss.

Selaginella has several advanced features in its life cycle: 1. differentiation of megasporophylls and microsporophylls; 2. megasporangia and microsporangia; 3. megaspores and microspores; 4. greatly reduced gametophytes with the male and female gametophytes differing in structure; 5. retention of the embryo sporophyte within the female gametophyte for at least some time; 6. arrangement of the sporophylls into cones. These may not all seem to be advances, or at least not important ones, but their significance should become more evident after we consider the life cycles of seed plants. It should be emphasized, however, that club mosses do not represent an evolutionary line between ferns and seed plants. Rather, all three groups probably represent divergent lines of evolution.

Pine. The pine serves as a representative of the gymnosperms. The pine tree is the sporophyte and bears its sporophylls in cones (Fig. 84). The large, woody cones that bear seeds are essentially stems with very short internodes and no continued apical growth bearing megasporophylls. The microsporophylls are borne in separate cones, these being much smaller than the cones of megasporophylls, non-woody, and of limited duration. Two microsporangia (pollen sacs) are borne on the underside of each microsporophyll. The numerous microspore mother cells undergo reduction division, each producing four haploid microspores. While still in the microsporangium the microspores develop into male gametophytes (the pollen grains). The male gametophyte of pine is even simpler than that of club mosses, initially consisting of only four cells. Two of these soon distintegrate, leaving only a generative cell (that later gives rise to two sperm) and a tube cell. The male gametophyte is retained within the microspore wall.

Each megasporophyll bears two megasporangia (ovules) on its upper surface, each containing only one megaspore mother cell. This undergoes reduction division with the production of four haploid megaspores, but three of these distintegrate and only one develops into a female gametophyte, which comes to occupy all of the space within the megasporangium walls when mature. It is composed of several thousand cells and has several archegonia at one end.

The male gametophytes (pollen grains) are widely

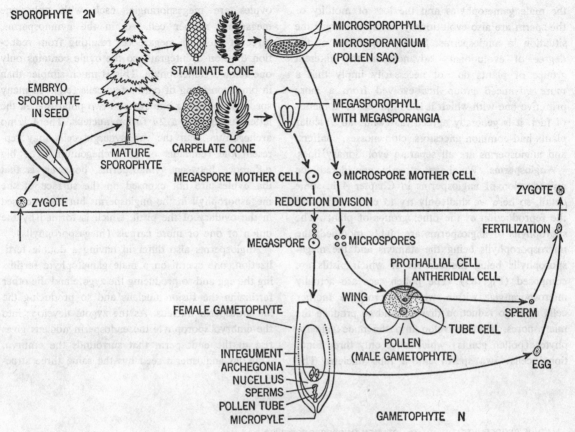

Fig. 84. Life cycle of a gymnosperm (pine).

distributed by the wind and some fall upon the megasporophylls near the megasporangia. The pollen then germinates, forming an elongated pollen tube that may grow into the megasporangium through a small pore known as the **micropyle.**

Once the pollen tube has reached the female gametophyte its tip ruptures and the sperm enter the archegonia and fertilize the eggs. Several zygotes may result, and each may develop into an embryo sporophyte. Unlike the sperm of the other species we have considered, the sperm of pine are not flagellated and so cannot swim, but the pollen tube brings them in close proximity to the eggs.

The ovules develop into the seeds of the pine. The integuments of the ovule become the seed coats, the female gametophyte becomes the endosperm of the seed, and the zygote grows into the embryo sporophyte of the seed. The seeds have wings that facilitate their dispersal by the wind.

It should now be evident that club mosses missed

having seeds by only a narrow margin. They have the embryo sporophyte embedded in the tissue of the female gametophyte, just as pine seeds do. However, this club moss structure falls to the ground from the megasporophyll before the embryo sporophyte is mature, and, since the megasporangium (and subsequently the female gametophyte) are not enclosed in integuments, there are no seed coats. The presence of seed coats is an essential characteristic of seeds, permitting the embryo to survive for some time through unfavorable conditions and providing for wider dispersal of the species. Seeds are one of the principal factors that have made gymnosperms and angiosperms so successful as the dominant land plants. In contrast, if conditions are not favorable for the growth of the embryo sporophyte of the club mosses at the time it falls to the ground, it will soon die.

Seeds represent the principal advance of pines over club mosses, but the still greater reduction of the male gametophytes and the loss of motility of the sperm are also evolutionary advances toward the situation in angiosperms. Such comparisons of the degree of evolutionary advances in the different groups of plants do not necessarily imply that a more advanced group has evolved from a more primitive one with which it is compared. As a matter of fact, it is generally agreed that while all vascular plants had common ancestors, club mosses, conifers, and angiosperms are all separate evolutionary lines.

Angiosperms. We have already considered the reproduction of angiosperms in Chapter 4 in some detail, so here we shall only try to correlate it with the reproduction of the other groups of plants. The sporophylls of angiosperms are highly modified, the microsporophylls being the stamens and the megasporophylls being the carpels of which pistils are composed (Fig. 85). The pollen sacs are actually microsporangia, within which microspore mother cells undergo reduction division and so produce the microspores. These develop into the male gametophytes (pollen grains) which have only three functional cells: two sperm and a tube nucleus. The

ovules are megasporangia, each with only one megaspore mother cell. As in the gymnosperms, three of the four megaspores resulting from reduction division disintegrate, so the ovule contains only one female gametophyte. This is much simpler than in pine, consisting of only a few cells (eight in many species), including one egg and two polar nuclei, the latter fusing into a $2n$ **fusion nucleus.** There is no archegonium, but the two synergid cells may represent the remnants of an archegonium. The big difference between gymnosperms, however, is that the ovules are not exposed on the surface of the megasporophyll in the angiosperms but are enclosed in the ovulary of the pistil, which is formed by the union of one or more carpels (megasporophylls).

Angiosperms also differ in having a double fertilization, one sperm on a male gametophyte fertilizing the egg and so producing the zygote and the other fertilizing the fusion nucleus and so producing the $3n$ endosperm nucleus. As the zygote develops into the embryo sporophyte the endosperm nucleus gives rise to the endosperm that surrounds the embryo. Thus an angiosperm seed has the same three struc-

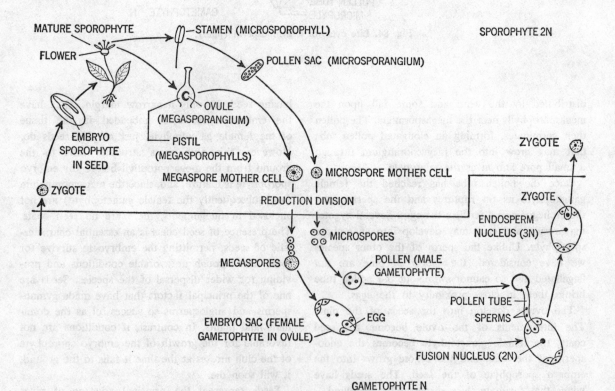

Fig. 85. Life cycle of an angiosperm.

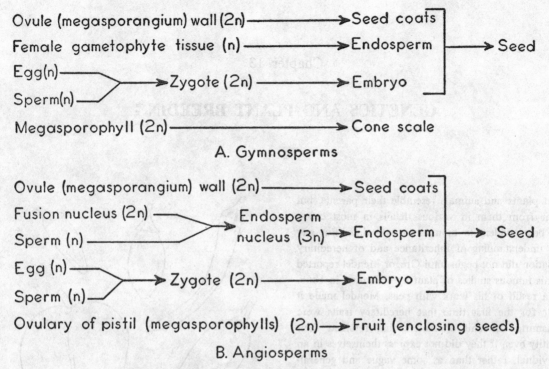

Fig. 86. Origin of seed and fruit structures.

tures as a gymnosperm seed (seed coats from the ovule wall, endosperm, and embryo), but the origin of their endosperms is different (Fig. 86).

As was noted in Chapter 4 the endosperm of some angiosperm species is all used up before maturity, and so the seed consists only of the embryo and the seed coats. While the ovules are developing into seeds the ovularies are developing into the surrounding fruits, still another important difference between angiosperms and gymnosperms. (The latter have no fruits because there are no ovularies.)

Evolutionary Trends in Reproduction. Certain long-range trends in the evolution of reproduction in the plant kingdom are evident:

1. There is a general trend toward increasing importance of the sporophyte generation, which is the dominant one in all vascular plants and essentially nonexistent (except for the zygote) in many algae.

2. There is a general trend toward decreasing importance of the gametophyte generation, which is the main plant in most algae and in the bryophytes and is reduced to the microscopic pollen grains (male gametophyte) and embryo sacs (female ga-

metophyte) of the angiosperms, with only a few cells each.

3. Isogamy in some algae and fungi gives way to heterogamy in the bryophytes and all higher plants.

4. The general trend toward production of eggs and sperm by separate individuals (males and females) rather than by one individual is not well established until the higher vascular plants are reached (the club mosses, gymnosperms, and angiosperms being our examples).

5. Advances in the complexity of sporophyte reproductive structures include sporophylls, heterospory (megaspores and microspores), increasingly greater differences between microsporophylls and megasporophylls, cones, seeds, flowers, and pistils and fruits, more or less in the order mentioned.

Yet, despite these evolutionary changes, the general pattern of plant life cycles is basically amazingly uniform throughout most plant groups: a haploid gametophyte generation producing gametes that fuse and form a zygote, and the beginning of the diploid sporophyte generation that produces spores by reduction division, the spores then developing into the gametophyte generation.

Chapter 13

GENETICS AND PLANT BREEDING

That plants and animals resemble their parents, but differ from them in various details in most cases, has been evident to man for a long time. But any real understanding of inheritance and of hereditary variation did not begin until Gregor Mendel reported on his famous studies of plant hybridization in 1866. As a result of his work with peas, Mendel made it clear for the first time that hereditary traits were transmitted as unit characters that retained their identity even if they did not express themselves in an individual, rather than as some vague and generalized blend of the total hereditary material of the species. Mendel's work made little impression on biologists until 1900, when three botanists independently pointed out its importance, and since then **genetics**—the scientific study of heredity—has made rapid advances.

The emphasis in genetics has been on hereditary variation, *i.e.*, in inheritable differences among the individuals of a species (Fig. 87), for heredity can be studied only when there are such differences to investigate. However, it should be recognized that many hereditary traits are constant and invariable generation after generation. It is such traits that taxonomists use in distinguishing one species from another and that make it possible to recognize white pines, for example, as a species distinct from longleaf pines, hemlock, white oaks, peas, or lima beans. Diverse as these species are, they all begin as zygotes that would be difficult to distinguish from one another under a microscope, but that contain genetic codes that cause growth and development to occur in a pattern characteristic of their species. Later in this chapter we shall summarize some of the things that recent research has revealed about these genetic codes and how they may control growth, development, and behavior of an individual.

Hereditary variation is usually the result of the reshuffling or recombination of the hereditary unit

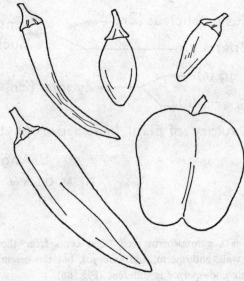

Fig. 87. An example of hereditary variation within a species. These are fruits of different varieties of peppers, *Capsicum annum*.

characters or potentialities (the genes) during the fertilizations and reduction divisions of sexual reproduction. More rarely, hereditary variations may result from **mutations**, *i.e.*, changes in the chemical nature of the genes or changes in chromosome structure or number. We shall consider each of these types of hereditary variation. It should be noted that individuals propagated vegetatively or by other means or asexual reproduction have exactly the same hereditary potentialities as their parent (except when mutations happen to occur), and are indeed just a separated piece of the parent. Thus, asexual reproduction makes for hereditary stability, while sexual reproduction makes for hereditary variation.

VARIATION BY RECOMBINATIONS

Monohybrid Differences. Each of the two sets of chromosomes in the cells of a diploid individual commonly contains a complete set of genes, and so there is a pair of any particular gene. When geneticists deal with only one such pair of genes at a time they are concerned with a **monohybrid difference.** For example, one of the gene pairs in peas that Mendel studied controls the height of the plants. If a plant contains two genes for tallness (TT) it will grow three feet or more high. If a plant contains two genes for dwarfness (tt) in its cells, it will grow only a foot or so high. A plant that contains one gene for tallness and one for dwarfness (Tt) will grow just as tall as a plant with two genes for tallness (TT). This is because tallness is **dominant** over dwarfness, the **recessive** gene of the pair.

If both genes of the pair present in all cells of the plant are identical (either TT or tt) the plant is **homozygous** for this trait, but if they are different the plant is **heterozygous** (Tt). A heterozygous tall plant and a homozygous tall plant look just the same, and so belong to the same **phenotype.** The dwarf plants belong to another phenotype. However, the homozygous tall plants and the heterozygous tall plants differ in their gene contents, and so belong to two different **genotypes.** Thus a phenotype consists of all plants that have the same visible trait, but the members of a phenotype may differ from one another in gene content and so may belong to more than one genotype. All dwarf plants, however, belong to one genotype (tt) as well as to one phenotype (dwarf).

Suppose we cross a homozygous tall pea plant with a homozygous dwarf plant by transferring pollen from one to the stigmas of the flowers of the other. All the offspring growing from the seeds thus produced will be tall plants. Whenever a homozygous dominant and a homozygous recessive are thus crossed (hybridized), they are referred to as the **parental** or P_1 **generation,** and the offspring are known as the **first filial** or F_1 **generation.** If members of the F_1 generation are crossed with one another the offspring constitute the F_2 **generation.** In our example the F_2 generation would consist of about ¾ tall plants and about ¼ dwarf plants, even though both parents were tall plants (Fig. 88). The F_2 phenotype ratio is thus said to be 3:1.

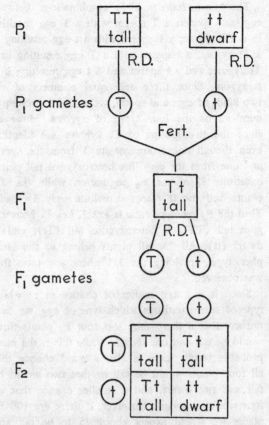

Fig. 88. A monohybrid with dominance. "R.D." indicates reduction division and "Fert." indicates fertilization. The checkerboard shows the possible combinations between F_1 gametes, *i.e.,* the possible fertilizations that result in the F_2 generation (the squares in the checkerboard). The trait is height of pea plants.

To understand what has been responsible for the phenotypes of the F_1 and F_2 plants we must consider what has happened during reduction division and fertilization. When the megaspore and microspore mother cells undergo reduction division the resulting spores contain only one set of chromosomes, and so only one gene of each pair. Since the tall P_1 plants (TT) contain only genes for tallness, all the spores, gametophytes, and gametes (eggs or sperm) contain one tall gene (T) per cell. Similarly, all the spores, gametophytes, and gametes of the dwarf P_1 plant (tt) contain one gene for dwarfness per cell. When the sperm from one parent fertilize the eggs from the other only one type of zygote can result—Tt. Thus, all F_1 plants are heterozygous tall. As a result of reduction division in an F_1 plant, half of the gametes will contain a gene for tallness (T) and half of them a gene for dwarfness (t).

There are four possible combinations between eggs and sperm: a T sperm with a T egg resulting in a TT zygote; a T sperm with a t egg resulting in a Tt zygote, a t sperm with a T egg resulting in a Tt zygote; and a t sperm and a t egg resulting in a tt zygote. Since there are equal numbers of the two kinds of eggs and sperm, we would expect equal numbers of the four types of zygotes. However, since the two groups of Tt zygotes are identical even though one group got its T from the sperm and one from the eggs, the heterozygous tall plants constitute ½ of the F_2 population while the TT plants and the tt plants constitute only ¼ each. Thus the F_2 genotype ratio is 1:2:1, *i.e.*, ¼ homozygous tall (TT), ½ heterozygous tall (Tt), and ¼ dwarf (tt). All the tall plants belong to the same phenotype, providing the 3:1 phenotype ratio that was observed.

Since it is just a matter of chance as to which type of sperm fertilizes which type of egg, we cannot say that if there were just four F_2 plants three would be tall and one dwarf. While this is the most probable result, there is also a good chance that all four plants would be tall or that two would be tall and two dwarf, and a smaller chance that all four would be dwarf. However, if there are 100 F_2 plants we would expect about 75 to be tall and about 25 to be dwarf, although we might be a little surprised if exactly this number of each resulted. The point is that we are dealing with the laws of chance or probability, and a rather large population is necessary if the theoretical ratios are to be realized. The larger the F_2 population, the closer we can expect the ratio to be to the theoretical 3:1.

Among the other hereditary traits of peas studied by Mendel were seed color (yellow dominant over green), flower color (red dominant over white), pod color (green dominant over yellow), and seed type (smooth dominant over wrinkled). Thousands of other examples of dominance have been found in other species of plants and animals. For example, in tomatoes smooth stems are dominant over hairy stems and tallness over dwarfness; in squash white fruits are dominant over yellow and disc shaped fruits over spherical fruits; and in corn presence of chlorophyll is dominant over absence (albino). In all these and other cases of dominance, all the F_1 plants show the dominant trait and in the F_2 ¾ show the dominant trait and ¼ the recessive trait.

However, in some gene pairs dominance is lacking and there is only **incomplete dominance** or blend-

ing. For example, in snapdragons a plant with the genotype RR has red flowers and a plant with the genotype rr has white flowers, but the heterozygous condition Rr results in pink flowers rather than red ones. Thus, if red snapdragons are crossed with white ones all the offspring are pink-flowering. In the F_2 generation ¼ of the plants have red flowers, ½ pink flowers, and ¼ white flowers (Fig. 89). The genotype ratio, as well as the phenotype ratio, is 1:2:1, since each different genotype is also a different phenotype.

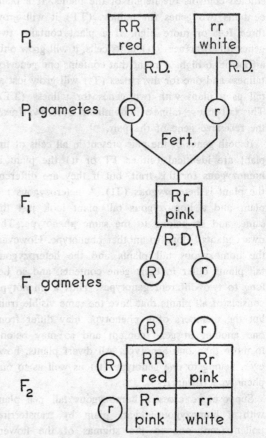

Fig. 89. A monohybrid with incomplete dominance (blending). The trait is flower color in snapdragons. In peas, red is dominant over white, so the F_1 would have red flowers, and in the F_2 there would be three red to one white.

Dihybrid Differences. Much more can be learned about the genes of an organism by following two pairs through a breeding experiment concurrently, *i.e.*, by investigating **dihybrid differences,** than by studying the two separately as monohybrid differences. As an example of a dihybrid we may take two seed characters of peas. If a pure line or homozygous plant producing yellow, smooth seeds (YYSS) is used as one P_1 and a plant homozygous for green, wrinkled seeds (yyss) as the other, all the F_1 plants will be heterozygous yellow and heterozygous smooth (YySs). The genes for seed color are located in a different pair of chromosomes from those for seed surface contour, and so they will assort independently when the spore mother cells of the F_1 plants undergo reduction division. That is, the chromosome containing Y may be on the same side of the equator at metaphase as the chromosome containing the gene S, which means that their mates containing genes y and s will be on the other side, but there is an equal chance that the chromosomes containing Y and s will be on one side and those

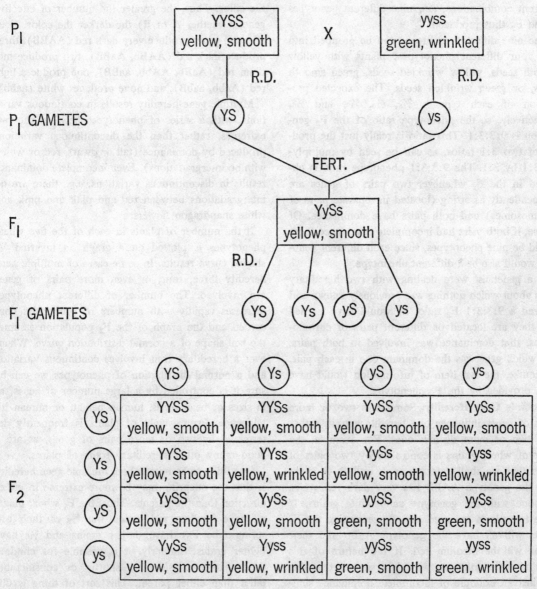

Fig. 90. A dihybrid, involving two characters of pea seeds: color and smoothness. The F_2 phenotype ratio is 9:3:3:1. Note, however, that if the F_2 phenotypes are tabulated by either color (green or yellow) or by smoothness (smooth or wrinkled) alone the phenotype ratio in either case is 3:1.

containing y and S on the other. Thus, considering the large number of spore mother cells produced by a plant, we can expect equal numbers of meiospores (and subsequently gametophytes and gametes) containing each of the four possible combinations of genes: YS, ys, Ys, and yS. Since there will be four types of eggs and four types of sperm there will be sixteen possible combinations in which they can fuse (Fig. 90), any one of the combinations being just as probable as any other. Analysis of the resulting genotypes, however, shows that there are only nine different combinations, *i.e.*, nine different genotypes in the F_2 thus produced.

The nine different genotypes can be grouped into only four different phenotypes: plants with yellow smooth seeds, yellow wrinkled seeds, green smooth seeds, or green wrinkled seeds. The expected proportion of each type is $\frac{9}{16}$, $\frac{3}{16}$, $\frac{3}{16}$, and $\frac{1}{16}$, respectively, so the phenotype ratio of the F_2 generation is 9:3:3:1. This ratio is really just the product of two 3:1 ratios, as can be seen by multiplying 3:1 by 3:1. The 9:3:3:1 phenotype ratio is obtained in the F_2 whenever two pairs of genes are independently assorting (located in separate pairs of chromosomes) and both pairs have dominance. Of course, if both pairs had incomplete dominance there would be nine phenotypes, since each different genotype would also be a different phenotype.

If a geneticist were dealing with two hereditary traits about which nothing was previously known and secured a 9:3:3:1 F_2 ratio, he would have learned that they are located on different pairs of chromosomes, that dominance was involved in both pairs, and which gene was the dominant one in each pair. Of course, the last item of information would have been provided by the F_1 phenotype.

Multiple Gene Heredity. Sometimes two or more pairs of genes influence one hereditary trait, rather than two or more distinct ones. For example, the color of wheat grains is controlled by two pairs of genes that we shall refer to as A and B. A plant with the genotype AABB has very dark red grains and one with the genotype aabb white grains. If these two plants are used as a P_1 generation the F_1 plants will all have the genotype AaBb and their grains will be medium red. If the nature of this heredity were unknown we might suspect that this was just an example of incomplete dominance similar to the flower color of snapdragons. However, when the F_2 generation is obtained the phenotype ratio is not 1:2:1 but rather 1:4:6:4:1, $\frac{1}{16}$ of the

plants having very dark red seeds, $\frac{4}{16}$ dark red, $\frac{6}{16}$ medium red, $\frac{4}{16}$ light red, and $\frac{1}{16}$ white (Fig. 91). The genotypes of the F_2 are just the same as in any dihybrid involving independent assortment, but the phenotypes are quite different because of a different type of gene interaction.

It is now obvious that we are dealing with a trait controlled by two pairs of independently assorting genes, each one of which has the same influence on grain color. Either A or B contributes toward a darker color, while neither a nor b contributes to the color. Thus, the greater the number of effective genes (whether A or B) the darker the color. Four effective genes produce very dark red (AABB), three produce dark red (AABb, AaBB), two produce medium red (AaBb, AAbb, aaBB), one produces light red (Aabb, aabB), and none produces white (aabb).

Multiple gene heredity results in continuous variations, with a series of phenotypes between the two extremes, rather than the discontinuous variations produced by dominance (tall or dwarf, red or white, with no intergradations). Even incomplete dominance results in discontinuous variations, *e.g.*, there are no intergradations between red and pink and pink and white snapdragon flowers.

If the number of plants in each of the five wheat phenotypes is plotted on a graph an inverted V-shaped curve results. In some cases of multiple gene heredity three, four, or even more pairs of genes are involved. The number of different phenotypes increases rapidly with numbers of gene pairs concerned and the graph of the F_2 population assumes the bell shape of a normal distribution curve. Whenever a hereditary trait involves continuous variation and a normal distribution of phenotypes we can be sure it is controlled by a large number of genes, as in sizes of bean seeds, human height, or human intelligence. Fruit weight of plants is frequently determined by two or more pairs of genes, as are a good many other hereditary traits of plants.

One thing to be noted about multiple gene heredity is that the offspring may be more extreme in either direction than the parents. Thus, the F_1 wheat plants all have medium red grains, but $\frac{5}{16}$ of their offspring (the F_2) have darker grains and $\frac{5}{16}$ have lighter grains. Similarly, it is possible for children to be either considerably shorter or considerably taller than either parent. This sort of thing is difficult to understand without a knowledge of how multiple gene heredity operates.

Linkage. Of course, each pair of chromosomes

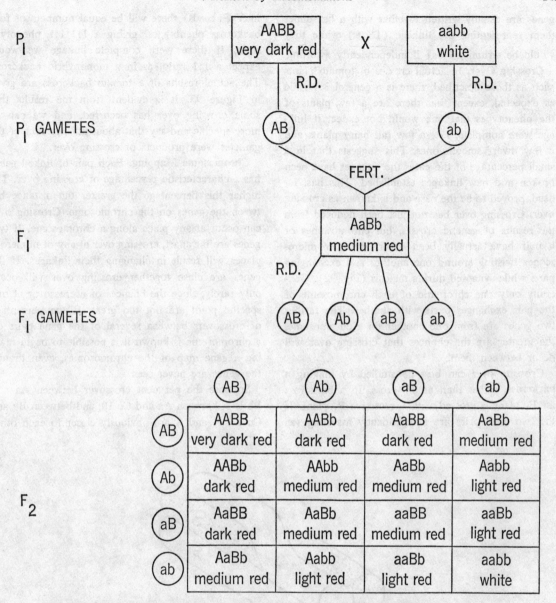

Fig. 91. Multiple factor heredity. A single trait (color of wheat grains) is controlled by two pairs of genes. Notice that the genes are redistributed as in an ordinary dihybrid, but the F_2 phenotype ratio (1:4:6:4:1) is quite different from that of a dihybrid with dominance.

contains many pairs of genes. Genes located in the same chromosome are said to be **linked** with one another and do not assort independently. In tomatoes the genes for height are linked with the genes for smooth or hairy stems. If a tall, smooth plant (TS TS) is crossed with a dwarf hairy plant (ts ts) the F_1 plants will be tall and smooth (TS ts), showing

that tallness and smoothness are dominant. However, there is no indication as to whether the genes are linked or independently assorting until the F_2 generation. If the genes are linked we would expect a ratio of three tall smooth to one dwarf hairy, rather than the 9:3:3:1 ratio for independently assorting genes. Thus, the F_2 phenotype ratios tell us whether two genes are linked or not. The linked genes are usually written together with a bar under them representing the linkage (TS ts) while they would be written (TtSs) if independently assorting.

Crossing Over. In actual crosses of tomato plants such as those described, there is a general 3:1 ratio as expected, except that there are a few plants of the phenotypes that one would not expect if linkage were complete, *i.e.*, a few tall hairy plants and a few dwarf smooth ones. This suggests that in a small percentage of the cases the linkages have been broken and new linkages established. This has, indeed, proved to be the case and is known as **crossing over.** Crossing over has not just been deduced from the results of genetic crosses, but chromosomes of a pair have actually been observed under microscopes twisting around one another and exchanging parts while synapsed during meiosis (Fig. 92). Generally only one chromatid of each chromosome of the pair exchanges parts with another. The farther two genes are from one another on a chromosome, the greater are the chances that crossing over will occur between them.

Crossing over can best be studied by making a **backcross,** rather than an F_2 cross. In a backcross an F_1 plant is crossed with a recessive P_1 plant. If the two gene pairs are independently assorting (as in YySs seeds) there will be equal numbers of four backcross phenotypes, giving a 1:1:1:1 phenotype ratio. If there were complete linkage we would expect a 1:1 ratio, as in a monohybrid backcross. The actual results of a tomato backcross are given in Figure 93. It is evident from the results that some crossing over has occurred, and we can be more specific and say that about 3.5 per cent of the gametes were products of crossing over.

Chromosome Mapping. Each pair of linked genes has a characteristic percentage of crossing over. The higher this percentage the greater the distance between the genes on the chromosome. Crossing over can occur at any place along a chromosome. If two genes are far apart, crossing over at any of numerous places will result in changing their linkages. If the genes are close together crossing over will occur only rarely, since the chances of a crossover at one specific point are not too great. If the percentage of crossovers between several of the gene pairs on a chromosome is known it is possible to begin making a gene map of the chromosomes, even though the genes are never seen.

Suppose the per cent crossover between Aa and Bb is 3, between Aa and Cc 10, and between Bb and Cc 13. A and B are obviously closer to each other

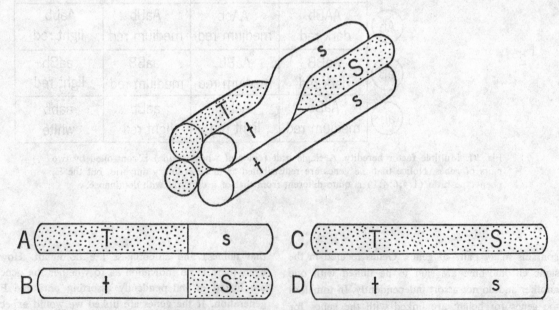

Fig. 92. TOP: The four chromatids of a pair of homologous chromosomes that are synapsed during meiosis. The two upper ones are twisted around each other and are about to exchange parts. BOTTOM: The four chromosomes received by the four meiospores resulting from the reduction division. **A** and **B** are the two chromosomes that have exchanged parts, while **C** and **D** have not. Note the change in gene linkages in **A** and **B**.

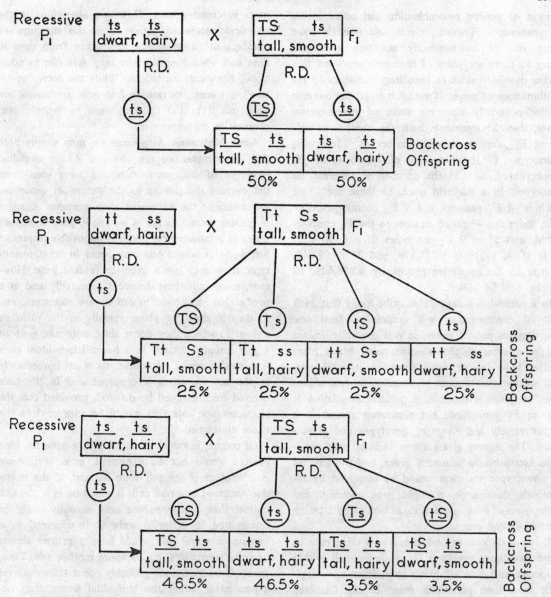

Fig. 93. Backcrosses of tall, smooth F_1 tomato plants with dwarf, hairy P_1 plants. TOP: Hypothetical results that would be expected if the genes were completely linked. CENTER: Hypothetical results that would be expected if the genes were on separate chromosomes and so independently assorting. BOTTOM: Results of an actual backcross. The small percentage of tall, hairy and dwarf, smooth phenotypes in the backcross offspring show that the genes are linked, but that some crossing over has occurred.

than either is to C. Furthermore, it is clear that the order on the chromosomes must be C, A, and B, since only this order would provide the proper relative distances between the three genes. After the per cent crossovers between any other gene on the chromosome is found, the new gene also can be

placed on the map. Extensive chromosome maps have been made of several plants and animals on which much genetic experimentation has been done, notably corn plants and fruit flies.

Extent of Variation by Recombination. We have seen that even in a dihybrid difference a considerable amount of genetic recombination can occur in an F_2 generation. However, plants and animals have many sets of independently assorting genes—as many as there are pairs of chromosomes—and this makes possible extensive hereditary variation by recombinations of genes. If we let n equal the number of independently assorting pairs of heterozygous genes, then 2^n represents both the number of different F_1 gametes and the number of different F_2 phenotypes, $(2^n)^2$ the number of squares in the F_2 checkerboard, and 3^n the number of different F_2 genotypes. In a dihybrid $n=2$, so there are 2^2 or 4 kinds of F_1 gametes and 4 F_2 phenotypes; further, there are 4^2 or 16 squares in the F_2 checkerboard, and 3^2 or 9 F_2 genotypes. If n is 10 then 2^n is 1,024, $(2^n)^2$ is 1,048,576, and 3^n is 59,049. If n is 23 the respective figures are 8,388,608, 70 trillion, and 94 billion.

In a natural population it is quite likely that each pair of chromosomes will contain at least one heterozygous pair of genes, so n is generally limited by the number of chromosome pairs. Many plant species have ten or more pairs, so it is obvious that an immense amount of variation by recombinations of genes is possible. A natural population is not an F_2 generation, but somewhat resembles it in the variety and range of genotypes and phenotypes. The figures given above include only variations ascribable to dominant genes, and the amount of phenotypic variation would be increased by incomplete dominance, multiple gene heredity, and other types of gene interaction, as well as by crossing over of linked genes.

It is not difficult to understand how there may be great hereditary variation by recombination within a species and how offspring may differ phenotypically from their parents in many details. Purebred lines of cultivated plants and domesticated animals have been bred until they are essentially homozygous for at least the desired traits, and so there is little variation by recombination. Also, natural populations that reproduce by some means of asexual reproduction show little hereditary variation.

VARIATION BY MUTATIONS

All the recombinations of genes that occur in the course of sexual reproduction do not alter the genes themselves, nor are the chromosomes altered except as the chromosomes of a pair may exchange parts in crossing over. However, alterations in the chemical nature of genes or in the structure of numbers of chromosomes may occur from time to time and when they do they may give rise to additional hereditary variations. While the term "mutation" is sometimes restricted to gene mutations, we shall use it in the broader sense to include also changes in chromosomes.

Gene Mutations. At present we may simply note that gene mutations are changes in the chemical structure of the genes. After we have considered the current theories as to the nature of genes we can consider the nature of such chemical changes in greater detail. As far as we know, gene mutations occur in a random way and it is impossible to predict just when or where one may occur or what phenotypic traits may result from the mutant gene. However, some mutations do occur repeatedly and at a predictable rate. Most mutations are recessives, and so do not show up phenotypically in the plant or animal in which they occur since only one gene of a pair usually mutates. In a haploid individual such a mutation would, of course, show up immediately in the cell in which it occurred and in the cells derived from this cell by division, provided that the character was one that would be expressed in the tissue concerned.

Of course, a mutation resulting in a gene for blue flowers would not be expressed, even if it were dominant or if the cell were haploid, if the mutation occurred in a root cell. A mutation in a plant or animal that can reproduce only sexually could be transmitted to offspring only if it occurred in a gamete or a cell that would have a gamete among its cell descendants, *e.g.*, a spore mother cell. Thus, many more mutations probably occur than are ever expressed in either the individual where they occurred or in its descendants. A recessive mutation might not show up for several generations after it appeared, when both the egg and sperm contributing to a zygote carried the mutant gene. Of course, this would be most likely to occur in matings of close relatives.

Dominant mutations occasionally occur in a cell giving rise to a bud, and so the branch derived from this bud will show the mutant character. This is known as a **bud mutation** or a **bud sport.** Such a bud mutation can be propagated vegetatively. Several of the valuable commercial varieties of fruits, such as the Golden Delicious apples, arose as bud mutations.

Mutations generally result in characteristics that are deleterious to the individual, however, and if they occur in cultivated plants or domesticated animals they are likely to be undesirable from an agricultural standpoint, too. The reason for this probably is that the hereditary traits of organisms have been subjected to long periods of natural selection (as in the case of domestic plants and animals) so the chances are that any change in their genes will be for the worse, since undesirable traits have been pretty well decimated by selection.

While ordinary environmental factors apparently do not induce mutations, and certainly do not induce mutations that will better adapt the individual to that specific environmental factor, certain kinds of radiation and some chemicals speed up the rate of mutation even though they probably do not cause any mutations that would not eventually occur otherwise. Mutations are promoted by ionizing radiation, including some wave lengths of ultraviolet light, gamma rays, X rays, and probably cosmic rays, as well as by ionizing particles such as alpha and beta particles emitted by radioactive substances. Much of the concern regarding radioactive fallout relates to the mutations it may promote, particularly in man, rather than to direct harmful effects on individuals exposed to it. However, radiation has been used to induce large numbers of mutations in peanuts and other crop plants in the hope that a few of the mutations will be for desirable traits. As a matter of fact, this hope has been realized and is leading to the production of valuable new varieties.

Once a new gene has arisen by mutation it can be transmitted from parent to offspring and can undergo recombination just like any other gene. Indeed, when we consider the matter from the long-range standpoint of evolution through millions and billions of years we can probably say that most if not all of the present-day genes are products of mutations that occurred at some more or less distant time in the past.

Chromosome Mutations. While chromosomes usually remain intact generation after generation, changes in chromosome structure do occur from time to time. Sometimes a piece of a chromosome becomes detached and lost during cell division. If the lost piece does not contain any genes essential for life hereditary changes may occur in individuals that eventually secure such a shortened chromosome from both parents, and even if the mate to the chromosome is the full length, recessive genes in the part missing in the shortened chromosome will be expressed, whereas they might otherwise be masked by dominants.

Chormosomes may also sometimes exchange parts with chromosomes of other pairs, quite a different situation from normal crossing over. Or, sometimes the central part of a chromosome or one end may become reversed or inverted, so there is a new order of the genes in the chromosome. In such chromosomal changes it might seem that there would be no change in heredity, since all the genes are still there. However, it has been found that genes may act differently when they have new gene neighbors (**position effect**), and so new phenotypes may result.

Polyploidy. As has been noted, sporophyte generation plants usually have the $2n$ or diploid number of chromosomes, *i.e.*, each cell contains two complete sets of chromosomes. However, a failure of reduction division to occur, or more commonly duplication of chromosomes in mitosis without the subsequent completion of cell division, may result in cells that have more than two sets of chromosomes. Such a condition is known as **polyploidy.** The most common type of a polyploid is a **tetraploid,** or a plant containing $4n$ chromosomes, but a triploid ($3n$) may arise by the union of a $2n$ gamete and an ordinary n gamete. Higher levels of polyploidy in plants are also known, *e.g.*, $5n$, $6n$, $7n$, $8n$, and even higher polyploids.

Polyploids are generally larger and more vigorous than the related diploids, probably because of the greater number of effective genes in multiple gene traits. The drug colchicine will induce polyploidy in plants by preventing the completion of cell division, and has been widely used both for experimental purposes and in producing new varieties of cultivated plants. Seed companies offer tetraploid varieties of many plants such as snapdragons and zinnias that they have created by the use of colchicine.

Many cultivated species differ from their wild ancestors in that they are polyploids. The commercial varieties of strawberries, for example, are high polyploids. Also, many wild species differ from other species in the same genus in being polyploids as well as in carrying different assortments of genes. Thus, it appears that polyploidy may have been an important factor in plant evolution. On the other hand, polyploidy appears to occur very rarely in animals.

Many polyploids, particularly triploids, are sterile

because reduction division cannot proceed normally and so they cannot reproduce sexually. If such plants can be propagated vegetatively (or can be produced anew each generation by crossing the parent lines) and have desirable traits, they may become widely cultivated. Many, but not all, of the sterile cultivated plants are polyploids.

THE NATURE OF GENES

It has been known for some time that chromosomes are composed primarily of very complex compounds known as **nucleoproteins,** but satisfactory theories as to the chemical nature of genes have not been proposed until recent years. It now appears that the nucleic acid component of the nucleoproteins, rather than the protein portion, constitutes the genes. The type of nucleic acid present in chromosomes and presumably constituting the genes is called deoxyribonucleic acid, or DNA for short. DNA molecules are apparently very long double coils made up of thousands of units linked together. Each unit consists of a molecule of de-

oxyribose (a 5-carbon sugar deficient in an atom of oxygen, $C_5H_{10}O_4$) with a phosphate group and an organic base attached to it. The organic bases, all ring compounds with nitrogen as well as carbon atoms in their rings and chemically similar to the alkaloids, are of four different kinds: adenine, thymine, cytosine, and guanine. For simplicity we shall refer to them hereafter as A, T, C, and G.

The two chains of the double coil DNA molecule are composed of deoxyribose (D) and phosphate (P) units linked together: D—P—D—P—D—P—D —P— etc. The two coils are linked together through the organic bases (Fig. 94). Because of their molecular structures, A can link only with T and C only with G. However, the four bases may occur in any possible order in the molecule, thus providing what amounts to a four-letter alphabet which serves as a genetic code. Since a DNA molecule (and probably a gene) consists of thousands of linked units, an almost unlimited variety of genetic information can be carried by the code. In any particular gene the sequence of the bases is always the same, this code providing a particular bit of genetic information, and even a single change in the sequence will provide

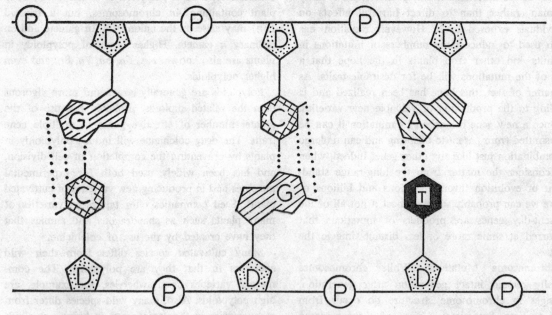

Fig. 94. A small portion of a DNA molecule showing the deoxyribose (D) and phosphate (P) units that make up the continuous part of each of the two coils, and the nitrogenous bases (A–adenine, C–cytosine, G–guanine, and T–thymine) that are attached to the deoxyribose units and link the two coils together by means of weak hydrogen bonds (dotted lines) between them. The shapes of the various components are roughly those of the actual molecules. Note the differences among A, C, G, and T that limit the bonding to A with T and C with G. Note also that a particular base, such as C, may be on either coil. The bases may occur in any sequence on the coil, thus providing a code.

new genetic information or perhaps no information at all, just as *blot* provides different information than *bolt* whereas *lbot* carries no information at all. Presumably a gene mutation is simply a change in the order of the organic bases, even a single change possibly being enough.

Apparently all organisms from bacteria to cabbages and protozoa to man have DNA genetic codes with the same A, T, C, G base, the marked differences between such diverse organisms as well as the smaller differences between each of them and their close relatives simply resulting from a different and characteristic sequence of these four organic bases in their DNA molecules.

The DNA molecules of a chromosome are apparently oriented across the chromosome and are presumably linked to one another by the protein component of the nucleoprotein. Thus, at the molecular level we might visualize a chromosome as being constructed roughly like a ladder, with the uprights representing the protein and the crossbars or rungs the DNA. However, the rungs may be zigzag rather than parallel or there may be only a single upright in the center.

When a chromosome duplicates itself in cell division, forming two chromatids, the duplication is perfect down to the exact sequence of the organic bases in the DNA molecules. This perfect duplication of the DNA molecules is presumably accomplished by the separation of the two coils of the molecule from one end and the building of a new coil against each of the two old coils from fresh deoxyribose-phosphate-base units. Since A can cross-link only with T and C only with G each old coil assembles a new companion coil identical with the one from which it separated. For example, if the sequence of bases in one coil started out ACTCGT-TAGC the sequence in its paired coil was TGAGCA-ATCG. After separation of the coils only a unit containing T could attach to the first position of the former, only G to the second position, an A to the third and so on. Thus, a perfect duplicate of the original mate coil is assembled.

GENE ACTION

There seems to be little doubt remaining but that genes exert their influences on the processes of organisms, and so in turn direct the pattern of growth, development, and behavior, by controlling the production of enzymes. Very few biochemical processes proceed without the catalytic activities of enzymes, and each of the many individual biochemical reactions of a plant or animal is activated by a specific enzyme. By controlling enzyme production genes determine just what enzymes the individual can produce, and so in turn what reactions can occur.

Whether a particular reaction *will* occur depends also on the environment. For example, a plant may have all the enzymes essential for chlorophyll synthesis but if an essential environmental factor such as light or magnesium is lacking chlorophyll cannot be synthesized. On the other hand, if even one of the essential enzymes is lacking chlorophyll cannot be synthesized even if all essential environmental factors are present. This is evidently the case in albino corn plants, their cc genotype not providing one of the essential enzymes. The dominant gene C permits synthesis of this enzyme, whether the plant is homozygous (CC) or heterozygous (Cc). Similarly, gene r fails to provide an enzyme essential for anthocyanin synthesis in sweet pea flowers and so the flowers are red. Structural traits are also controlled through the gene to enzyme relationship. At least some kinds of genetic dwarf plants are dwarfs because they produce little or no gibberellin. If they are sprayed with gibberellin they grow as tall as the genetically tall plants and are phenotypically indistinguishable from them. Apparently gene T provides some enzyme essential for gibberellin synthesis, while gene t does not.

Much of the information we have about the relationships between specific genes and specific enzymes has been obtained from experiments with the orange bread mold, *Neurospora crassa*. Many biochemical mutants unable to synthesize one of the essential amino acids or vitamins usually produced by the mold have been induced by exposing the mold to radiation, and the mutant genes have been identified. These biochemical mutants will grow only when provided with the missing vitamin or amino acid.

Although the control of enzyme production by genes seems to be well established, the precise way in which this takes place is still mostly theoretical. However, present knowledge does permit the formulation of an attractive and plausible general theory. The DNA of the genes apparently controls the synthesis of another kind of nucleic acid (ribonucleic acid, or RNA) and transfers its coded genetic information to the RNA. RNA differs from DNA in that the sugar component is ribose and not deoxyribose, and in place of thymine another organic base (uracil) is present. Also, the RNA molecules are apparently not double coils like the DNA molecules.

Once produced, the RNA apparently goes from the nucleus into the cytoplasm where it is present on small spherical structures known as **microsomes** (or **ribosomes**) that are a part of the **endoplasmic reticulum.** The microsomes are so small that they can be detected only with electron microscopes. Proteins are synthesized on the microsomes, including enzymes as well as other proteins. The genetic code transferred from the DNA to the RNA controls the sequence in which just the right kinds of amino acids are put together, thus producing a specific and characteristic protein.

Since the four-unit DNA and RNA codes must be able to code for the twenty different amino acids that make up proteins, it is obvious that any one of them (*e.g.,* A or C) cannot code for an amino acid. Even in pairs of two there can be only sixteen possible combinations, so they must at least operate in groups of three (triplets) to provide the necessary number of codes. However, all possible combinations of four units into triplets add up to 64, which is far more than needed to code for twenty amino acids. Abundant experimental evidence is now available that the DNA and RNA codes actually are triplets, and also that several different triplets may code for one amino acid (although any particular triplet can code for only one specific amino acid). For example, the RNA codes UUU and UUC code for the amino acid phenylalanine; UCU, UCC, UCA and UCG for serine; and AAA and AAG for lysine. Two triplets (UAA and UAG) do not code for any amino acid, but apparently function like a period to signal the end of a protein chain of amino acids. Thus, all 64 possible triplets provide code information.

PLANT BREEDING

The development of modern genetics has not only provided much important basic biological knowledge, but it has also been of immense practical value in the breeding of improved varieties of cultivated plants and domesticated animals. The results have included larger and more attractive flowers, increased yields of both farm and garden crops, better quality, and, perhaps most important of all, disease-resistant varieties of high quality.

Plant breeding was practiced long before Mendel laid the foundation for modern genetics, and some desirable results were obtained, but since the plant breeders knew practically nothing about the transmission of hereditary traits they worked largely by trial and error and spent much more time and effort than would have been required had they known the scientific facts about heredity. Luther Burbank was the last of the important non-scientific plant breeders, and even though much genetic knowledge had accumulated by the time he did his work, he made little use of it. Plant breeding, or at least the selection of the best available plants for propagation, undoubtedly went on long before the dawn of written history since the crop plants cultivated by primitive peoples were in most cases quite different from all of their nearest wild relatives.

The modern scientific plant breeder makes use of all available genetic knowledge and techniques. He determines the genotypes as well as the phenotypes of the desirable traits of parental stocks of several different varieties that he wishes to cross so as to incorporate into one variety the desirable traits of several varieties, at the same time eliminating undesirable traits. Knowledge of genetic principles and the genotypes of the plants greatly facilitates and speeds the accomplishment of the desired results, although much time and effort are still required. In addition to recombining genes, the plant breeder may induce mutations by the use of radiation or chemicals or induce polyploidy by the use of colchicine.

Once a desirable new variety is obtained the problem of propagating it in commercial quantities remains. Most varieties of farm and garden crops propagated by seeds are pure lines, *i.e.,* they are homozygous for the desired traits and so will breed true for generation after generation with little or no hereditary variation as long as they are not cross-pollinated with pollen from another variety. It generally takes some time to establish such pure lines. In trees, shrubs, and other plants that can readily be propagated vegetatively the time-consuming task of establishing pure lines is not necessary. Once a desirable individual is obtained as a result of gene recombinations, mutations, or polyploidy it can be propagated vegetatively and will have offspring identical with it with respect to hereditary characteristics even though it is extremely heterozygous. This is an advantage animal breeders do not have. If the seeds from a certain variety of apples are planted it is likely that none of the resulting trees will bear apples like those of the parent tree and most will have quite inferior fruit (though a few may be desirable new varieties), all because of the extreme heterozygosity and wide range of hereditary variation found in apples. However, the parent variety can be propagated true to form indefinitely by grafting.

The development of hybrid corn, which was first planted on any scale as a crop in 1933, introduced a new approach to plant breeding. At present some 95 per cent of the corn raised by farmers is hybrid corn, which not only has much higher yields than the old varieties but also is uniform in many respects: ear size, plant height, and quality of the grain. Hybrids of other crop species such as tobacco are now also available, though none of these hybrids is used as extensively as hybrid corn.

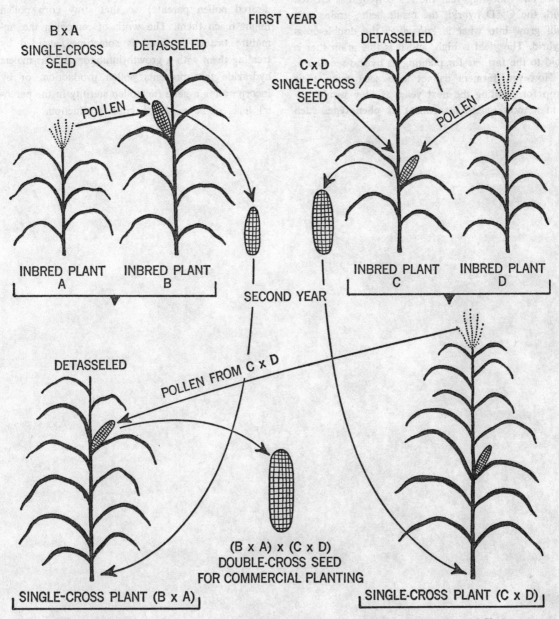

Fig. 95. Diagram showing method of producing double-cross hybrid corn. (*Courtesy of U. S. Department of Agriculture.*)

In producing hybrid corn, four pure line parental strains are developed, each of them with both desirable and undesirable traits. Generally they are low-yielding, produce small ears, and are quite unsuitable for planting by farmers. These pure lines are produced by inbreeding and are maintained year after year to provide continued parental stocks. To produce hybrid corn, A is crossed with line B and line C with line D, thus giving two true hybrid or F_1 generations (Fig. 95). The yield of corn is quite low. The following year the A×B hybrid is crossed with the C×D hybrid, the result being grains that will grow into what is known as the double-cross hybrid. The yield is high, and it is this grain that is sold to the farmers for planting as hybrid seed corn.

However, farmers cannot save seed from their crop for planting the next year, as they would obtain a most varied assortment of phenotypes such as would make up an F_2 generation. The farmers are thus dependent on the hybrid seed companies for seed corn year after year. The seed corn is expensive, because of the skill and labor involved in producing it, but the high yields still make it profitable for farmers to buy hybrid seed rather than one of the older varieties. In making the crosses it is essential that the line bearing the ears be detasseled so it will not be self-pollinating and also that the plants not be too near any other corn (except the desired pollen parents) so that stray corn pollen might reach them. The work of removing the immature tassels by hand is sometimes avoided by treating them with a growth inhibitor such as maleic hydrazide that prevents pollen production, or by incorporating a gene for pollen sterility in the parental lines to be used for grain production.

Chapter 14

THE EVOLUTION OF PLANTS

Discussions of evolution commonly center upon the evolution of man, or perhaps the evolution of animals in general, but it is clear that the numerous species of plants that inhabit the earth are also products of evolution. There are two basic questions regarding evolution: 1. Did the million or more species of plants and animals that inhabit the earth originate by evolution or in some other way? 2. If evolution is the method of origin of species, how has it occurred?

Biologists have long been in agreement that species have arisen by evolution, but there has been considerable disagreement as to the way in which evolution has occurred. Various theories have been proposed, but most of them have been found wanting. At the present time there seems to be general agreement that Darwin's theory of natural selection is the only theory in accord with the known facts, but there is still disagreement as to details, and various modifications have been proposed in an effort to provide a theory in accord with all current biological data and knowledge.

EVIDENCE FOR EVOLUTION

The only alternative to the origin of species by evolution is that each of the many species originated or was created independently in its present form from non-living matter, either at a single time in the past or from time to time. The former hypothesis is completely contradicted by all available scientific evidence, particularly the fossil record of the past life of the earth. While the latter could be fitted into the fossil record better it is not in accord with other known facts, and it is certainly less plausible that a complex organism would suddenly arise from non-living matter (or perhaps from a completely unrelated living thing) than that it would arise from a closely related ancestral species by relatively minor evolutionary changes in hereditary traits. Yet for many years biologists as well as laymen considered that at least some species could and did arise by such spontaneous generation. In ancient and medieval times it was commonly believed that leaves of trees falling into water might become fish while those falling to the ground might become birds, that goose barnacles might turn into geese, that piles of old rags could become mice, and that decaying meat turned into maggots.

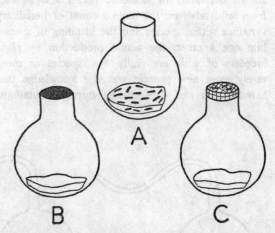

Fig. 96. Redi's experiment on spontaneous generation. Pieces of meat were placed in three flasks, Flask A being left open, Flask B being tightly closed, and Flask C being covered tightly with cloth. Maggots developed in the meat in Flask A. That this was not spontaneous generation was shown by the absence of maggots in the meat in Flasks B and C, since flies that laid the eggs that developed into the maggots (larvae) could not reach the meat. However, flies attracted to Flask C by the odor of the meat laid eggs on the cloth and maggots developed there. Redi had previously found that there was a connection between the flies and maggots. The meat in all three flasks decayed. Redi repeated this experiment many times and also used dead snakes and fish instead of meat.

About 1650 Francesco Redi conducted a simple but conclusive experiment that showed that decaying meat does not contain maggots if screened from the flies that laid their eggs in the meat, the maggots being the larvae of the flies (Fig. 96). Scientists, though by no means all laymen, then began giving up their theories of spontaneous generation. However, when bacteria and other microorganisms were discovered not too much later many biologists believed that these tiny plants and animals arose by spontaneous generation, even though larger plants and animals did not. The spontaneous generation of microorganisms was not finally and fully disproved until Louis Pasteur conducted his famous experiment in 1864 (Fig. 97).

The evidences for evolution come from a variety of sources, principally from fossils; the presence of comparable (homologous) structures in different species more or less closely related to one another; physiological and biochemical similarities among related groups of species; the geographical distribution of species; similarities in embryological development of species quite diverse as adults and embryological stages that resemble adults of simpler forms; the known derivation of domestic plants and animals from wild relatives; the large amount of hereditary variation within species and the blending of species into one another; the actual production by plant breeders of what are really new species or more rarely even new genera; and the knowledge that heredity can and does change through mutations.

We shall discuss these various types of evidence briefly.

Fossils. Sedimentary rocks (those that have been formed from deposits of mud, sand, calcium, carbonate, or other materials over millions of years) commonly contain fossils of plants (Fig. 98) and animals. These deposits commonly occurred at the bottoms of ancient oceans, lakes, bogs, or swamps and may later have been raised to become dry land, perhaps even parts of mountains. Geologists can date the age of the various sedimentary rocks, and so the fossils provide us with considerable information about the kinds of plants and animals that lived at various times in the geological past. Later in this chapter we shall consider the past plant life of the earth. Fossils may be either the actual remains of the organism more or less highly modified and sometimes changed into rock (petrifaction), or they may simply be imprints of the organism in the surrounding deposits.

The record of plant fossils is far from a complete record of the past plant life of the earth, however. Some plants, particularly most of the nonvascular plants, are so soft and delicate and subject to such rapid decay that they are very unlikely to leave fossil remains. Even species that might leave fossils if promptly covered by deposits or at least water that would retard decay may undergo complete decay in the air. Also, some of the older sedimentary rocks have been so compressed and distorted that any fossils they contained would be hard

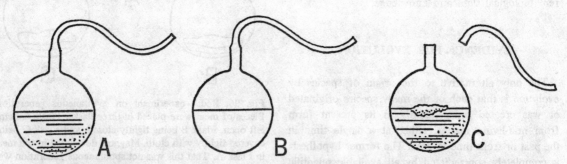

Fig. 97. Pasteur's experiment on spontaneous generation. A–Many bacteria grew in nutrient broth in a flask when the broth was not sterilized by heating. B–When broth was sterilized by heating after being placed in a similar flask no bacteria appeared and the broth remained clear, thus showing that bacteria could not arise from the broth by spontaneous generation. C–When the neck was removed from Flask B bacteria from the air fell into the broth and grew, showing that the heating did not destroy the capacity of the broth for nourishing bacteria. The long S-shaped necks permitted entry of air into the flasks without entry of bacteria, and were used because previous experiments using sealed flasks were criticized on the basis that some essential component of the air had been destroyed by the heating.

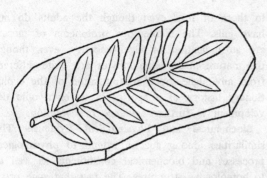

Fig. 98. The leaf of a fern, a plant fossil.

to recognize, while others have been largely or entirely eroded during subsequent geological ages. Other sedimentary rocks are still below the ocean and relatively inaccessible. However, the record of fossil plants that we do have makes it quite clear that many species that once inhabited the earth are now extinct and that as late as a hundred million years ago most of our present species of plants were not in existence, while many of them appeared for the first time much later than this. Also, the fossil record makes it clear that the older the sedimentary rocks, the more primitive the most highly developed type of life was.

Geographic Distribution. Most native species and genera of plants, and more rarely entire families, have a limited geographical distribution, rather than being present wherever the environment is favorable for their growth (as might be expected if evolution had not occurred). It is generally possible to identify a center of dispersal where the ancestral species apparently evolved, where the number of present-day species is most numerous, and from which the descendant species fanned out into other regions suitable for their growth. The spreading from a center of distribution may be restricted by geographical barriers such as mountain ranges, oceans, or other environments that may be unfavorable to a particular organism.

Sometimes two related species or two groups of related species are found in two regions separated by a major barrier such as an ocean, but in such cases there was no barrier in the geological past, at the time that the species had evolved. For example, many species, or at least closely related species, of trees are found only in eastern Asia and western North America and evidently migrated from one region to another over a former land bridge that disappeared millions of years ago. Most oceanic islands have their own characteristic species of plants and animals that are found no place else. They evidently evolved on the islands, but generally their closest relatives are found on the nearest mainland. The strange animals of Australia are quite well known, but the plants are also quite different from those of other continents.

Homologous Structures. The presence of comparable or homologous structures in species that are more or less closely related indicates common ancestry, and is difficult to explain logically on any other basis. The fact that the fruits of all oaks are acorns and that no other plants produce acorns is one of the structural features that makes it obvious that all species of oaks are so closely related that they should be grouped in a single genus. The relationship of the various oaks to one another was obvious to laymen, who applied the common generic name "oak" to the group even before botanists set up the genus *Quercus*. Wherever there is any real relationship there must be common ancestry.

Many structural features are common to entire orders, families, classes, or phyla, rather than just to more limited groups such as species or genera. For example, all members of the composite or sunflower family have their flowers borne in compact heads, all members of the angiosperm class have fruits in contrast with the lack of them in gymnosperms, and all members of the phylum *Tracheophyta* have vascular tissue and almost all have roots, stems, and leaves, none of these being present in the other phyla of plants. It seems evident that the larger the group having a structure in common, the earlier this structure arose in evolution. The fact that the various specialized structures of vascular plants, such as spines, thorns, and tendrils are obviously just modifications of the basic organs such as stems or leaves, provides additional evidence for evolution.

It should be noted that evolution may result in simplification, reduction, or even elimination of a part as well as in the greater development of a structure. This sort of trend has been particularly marked in the reduction of the gametophyte generation as evolution has proceeded. Flowers provide an excellent example of how fundamental structural patterns may be greatly modified in detail by evolution, yet with all the immense variation in details of flower structure from one species of angiosperm to another, and even with the deletion of some parts such as petals or sepals from the flowers of some species, flowers are all still built on a basic structural pattern common to all angiosperms. Furthermore, the stamens and pistils of flowers are recognizable as modified sporophylls and so flowers are homologous with the cones of gymnosperms, club mosses, and horsetails and the stamens and the carpels that make up pistils are even homologous with the sporophylls of the ferns. Such long-range modifications of the three basic organs of vascular plants—leaves, stems, and roots—are difficult to explain on any other basis than common relationship in the more or less distant past, and evolution.

Biologists must be careful, however, to use truly homologous structures and not analogous ones in establishing relationships and plotting the course of evolution. Analogous structures are similar in function but not in basic structure. Thus, both butterflies and birds have wings and use them in flying but there are no basic structural similarities between the wings of the two at all. Bird wings are, however, homologous with the arms of man or the front legs of a dog and have many similarities in bone and muscle structure though they look quite different externally and have different functions. The so-called leaves, stems, and rhizoids of moss gametophytes are analogous with the leaves, stems, and roots of vascular plant sporophytes, but their basic structure is quite different and so they are not homologous and do not provide a basis for establishing a close relationship.

Care must also be taken in not mistaking examples of parallel evolution as evidence for close relationship. Thus, the fact that some club mosses have almost evolved seeds does not mean that they are the ancestors of our seed plants, but rather that they constitute another evolutionary line. In Africa there are many members of the spurge family present in deserts that look much like a number of species of cactus growing in our southwestern deserts. However, the two families are not closely related, and have quite different basic structures, particularly with respect to types of flowers. The evident similarities are only superficial and not basic, and are simply the result of similar lines of evolution in similar environments of two quite unrelated groups.

Even when the structures of adults of two groups are quite different, similarities in early development may indicate a more or less distant relationship. The embryos of mammals have gill slits, quite similar to those of fish, even though the adults do not have gills. The filamentous protonema of mosses are an indication of algal ancestry, even though the mature moss gametophytes are quite different from any algae. Such recapitulation of the evolutionary history of a species in its embryonic development is further evidence of evolution.

Biochemical and Physiological Evidence. The similarities among species extend to physiological processes and biochemical constituents as well as to homologous structures. The fact that such processes as respiration and cell division occur in an essentially similar way in both plants and animals indicates some degree of relationship among all living things, just as much as the general similarities in basic cell structure do. Also, photosynthesis apparently proceeds in the same general way in all higher plants and also at least in green algae, although the photosynthetic bacteria have evolved somewhat different patterns. The fact that green algae, bryophytes, and vascular plants all have the same chlorophylls and carotenoids, while other algae and bacteria have different assortments of photosynthetic pigments, provides evidence for relationship among the former groups. On a much more limited scale, the production of alkaloids principally by members of certain families of plants, such as the nightshade (potato) and poppy families, indicates the evolution of certain metabolic patterns within these families and not in other related families.

One of the best biochemical evidences of evolution and relationship is the fact that the proteins of two closely related species are much more similar to one another than those of more distantly related species. An animal such as a rabbit sensitized to proteins from a sugar maple tree, for example, will react violently to subsequent injections of protein from this species, somewhat less violently to proteins from other species of maple, and not at all to proteins from unrelated species such as pines or potatoes. Samples of serum from such a sensitized animal will form precipitates proportional to the degree of similarity in the proteins, and such serum precipitation tests have been used in checking on the degree of relationship between various species of plants, as well as various species of animals. Other methods of analyzing for specific proteins, such as electrophoresis, can also be used. Also, the nucleic acids of related species are more similar than those of distantly related species.

Genetic Evidence. Some of the best evidence for evolution, or perhaps more precisely evidence that evolution is a plausible theory and without doubt a correct one, comes from the science of genetics. After all, evolution results from a series of changes in hereditary traits and so genetics provides the basic facts needed for an understanding of the process of evolution. The fact that mutations can and do occur provides the changes in heredity and the appearance of new hereditary potentialities essential to evolution. Since life has been present on the earth for at least three billion years, time is not a barrier to the occurrence of the numerous mutations that would be necessary in the evolution of the highest vascular plants from the simplest and most primitive plants. However, gene mutations are not the only source of the hereditary variations essential for evolution. Chromosomal mutations and polyploidy also provide new phenotypes, and even the variation provided by recombinations of genes may be of evolutionary significance, as we shall see later. This is particularly true when related species hybridize, as is particularly common in some genera such as the oaks and hawthorns, some of the hybrids being distinct enough that they really constitute new species. All types of hereditary variation can contribute toward evolution, and if there were no hereditary variation of any sort there would be no evolution.

Cultivated plants provide particularly good examples of evolution at a rather rapid rate. Most cultivated species are quite different from their nearest wild relatives and the evidence is that they never existed as wild species in their present forms but have evolved from their wild ancestors during the relatively brief period that man has cultivated plants, largely as a result of the selection of particularly desirable phenotypes for propagation. Most of this selection occurred among primitive peoples before the time of written history. During the brief period of less than a century that heredity has been studied scientifically geneticists and plant breeders have already produced many new varieties of plants, some that are distinct enough that they are really new species, and even a few that constitute new genera. One of the latter was produced by crossing radish (*Raphanus*) with cabbage (*Brassica*), two genera of the mustard family. A resulting hybrid that bred true to form was so different from either parent that it was placed in a separate new genus, *Raphanobrassica*. Though of great scientific interest it is use-less economically, since it has the roots of cabbage and the leaves of radish.

THE METHOD OF EVOLUTION

There is no doubt among biologists but that species have originated by evolution, but there has been considerable doubt and controversy about the way in which evolution has operated. Two principal theories as to the mechanism of evolution have been proposed: Lamarck's Theory of Inheritance of Acquired Characteristics in 1809 and Darwin's Theory of Natural Selection in 1859. We shall consider each of these briefly.

Inheritance of Acquired Characteristics. Lamarck proposed that acquired characteristics are transmitted to offspring and so the use or disuse of a particular organ would determine the course of its evolution. For example, hoofed animals like horses are really walking on the greatly developed nail of one toe, the other toes being rudiments, but their ancestors had toes like other mammals. Lamarck thought this resulted from the ancestor running rapidly on tiptoe, so the used toe developed greatly while the unused ones atrophied, and that after generations of this the present-day hoof evolved. Similarly, the long neck of the giraffe would result from the short-necked ancestors stretching their necks to reach leaves high in trees, acquiring longer and longer necks generation after generation. The concept of use and disuse is harder to apply to plants than animals, but as we have seen in Chapter 9, environment may have marked influences on the structure of plants. Some species that are hairy when growing on a high mountain are smooth in the lowlands. Lamarck would consider that such environmental variations would be transmitted to offspring, and so give rise to new species. Desert plants would evolve because of a direct influence of the dry environment on the plants, and such changes would then be transmitted to the offspring.

Lamarck's theory would be particularly effective in explaining the fact that organisms are adapted to the environment in which they live naturally, for the environment would play a direct role in initiating hereditary variations that would adapt the organism to its environment. However, the theory has been found to be quite unacceptable by biologists for several reasons, principally because no one has ever found any evidence that acquired traits (environmental variations) are ever inherited. On the contrary, there is much evidence that they are not inherited and do not change the heredity of the individual in any way. For example, the seeds of the hairy mountain plants, just noted, give rise to smooth plants when planted in the lowlands. It is common knowledge that a son does not inherit well-developed muscles from his athlete father and that the son of a famous pianist does not inherit the skill in playing the piano that his father acquired by much practice.

Yet, although the inheritance of acquired characteristics has been discredited scientifically, popular discussions of evolution often are based on this theory. Thus, we read that the human appendix has become small and useless because humans have not eaten much roughage, and that a rabbit's appendix is well developed because it does. Predictions are made that in the future human arms and legs will become smaller and more poorly developed because we use them less and less, while heads will become larger because we use our brains more and more. Both explanations are strictly Lamarckian, and not at all in accord with biological facts.

We now know that evolution occurs as a result of mutations, recombinations, and other genetic variations and that the environment does not cause such hereditary changes to arise from environmental variations. Heredity and environment both influence the internal processes and so in turn growth and development, but do not directly interact with one another. Of course, ionizing radiation does increase the rate of mutations, though it apparently does not cause mutations that would not occur otherwise eventually. This is quite a different thing, however, than the inheritance of acquired traits brought about by the action of the environment. For one thing, mutant traits are rarely expressed in the individual in which they occur and for another mutations are apparently random and in no way adaptive to the environmental factor that may have caused them.

Natural Selection. Darwin's Theory of Natural Selection holds that 1. there is a great deal of hereditary variation within most species; 2. there are many more individuals of any species produced than the available sites can accommodate; 3. intense competition between members of the species and with members of other species results, those best adapted to the environment surviving and those not well fitted to the environment dying out; 4. those individuals surviving this natural selection will pro-

duce the next generation and so transmit their desirable hereditary traits, while the less desirable traits of the individuals that did not survive to reproduce are lost.

As a result of such natural selection a species living in any particular environment would, after a relatively few generations, become quite well adapted to that environment and it should become quite well stabilized phenotypically. Any mutations that would occur, or any recessive traits that would be uncovered by recombinations, would be likely to make the individual less fitted for survival and so these traits would probably be lost while the established ones would be transmitted from generation to generation. There would be some chance, of course, of new traits that would make the individual better adapted to the environment and these would then be transmitted to offspring, and so the species might gradually change in character, and might eventually become so different that it would be classed as a new species that had replaced the ancestral one.

However, evolution would be more likely to occur if there were a major shift in environment, as occurs from time to time. Under these conditions it is highly probable that mutations and new recombinations of genes would result in new phenotypes better adapted to the new environment than the older ones. Such a change in environment would thus greatly speed up the rate of evolution and would probably result in the replacement of the ancestral species by one or more new ones. Of course, the change in environment could be so drastic that no member of the species could survive and so the species would become extinct. This has happened many times in the past. Environmental changes include not only changes in the physical environment but also the appearance of other species not previously present in the community either by evolution or invasion from other areas. At least some of the established species of the community might not be able to compete successfully with these new additions (particularly if they happen to be pests or parasites). The result might be either extinction or the natural selection of new phenotypes able to survive, and so evolution.

Another situation favorable to evolution is provided when a species spreads from its original habitat to other and somewhat different environments. Populations with different phenotypes may then appear in each of the environments as a result of natural selection. These different populations may merge into one another by intergrading forms and may be able to interbreed with one another freely, and so would simply be regarded as varieties of a single species. However, if the differences include marked enough changes in reproductive structures or processes so that the several populations can no longer interbreed freely, or if the populations become geographically isolated from one another, intergrading forms will disappear and the differences that make each population adapted to its particular environment will become fixed and accentuated by continuing natural selection. Thus, one to several new species may arise while the ancestral species continues to flourish in the original habitat. Over long periods of time and large areas the differences in populations brought about by natural selection in these ways may become marked enough that new genera, families, orders, classes, or even phyla may evolve, as well as new species.

While long periods of time are probably usually involved in the evolution of new species, let alone the larger groups, evolution of new varieties or strains may occur after just a few years of natural selection. We shall give only two of many possible examples, both influenced by man's intervention and so in a sense artificial selection rather than natural selection.

When antibiotics were first discovered they promised to wipe out many species of harmful bacteria such as the *Staphylococcus* that cause boils and other infections. However, after a time it was found that strains of these bacteria no longer controlled by a certain antibiotic were appearing. They were either mutants or phenotypes that had constituted only a small fraction of the original population, but when the great bulk of the population was eliminated by the antibiotic these resistant strains took over and became the ancestors of the new population. Some strains have now become resistant to most of the common antibiotics and are particularly abundant around hospitals, where antibiotics are used extensively. Similarly, there are now populations of flies resistant to DDT and other insecticides, and these are also products of selection.

In Chapter 10 we pointed out that the rabbits which were introduced into Australia and became pests were greatly reduced in number by infecting them with a fatal rabbit disease, but that now the few individuals immune to the disease are repopulating the country. This is another example of natural

selection. In England there is a species of moth that includes both light- and dark-colored phenotypes. In the past the dark-colored phenotypes were rare because when they rested on the light-colored bark of trees they could readily be found by birds that eat them. However, around industrial cities where the tree trunks are darkened by soot it is the dark-colored phenotypes that survive and the light-colored ones that are easily found by birds. The result is two distinct varieties, one living around cities and one in rural regions. The resulting spatial isolation reduces the chances of interbreeding of the two varieties, and may eventually result in other differences great enough to make them two distinct species.

Since Darwin knew nothing about modern genetics, and was apparently even quite unaware of the importance of Mendel's work, he could not provide any valid suggestions as to the nature of the hereditary variations that are subject to natural selection. However, we now know that gene mutations, chromosomal mutations, polyploidy, and even just recombinations of genes all provide hereditary variations subject to natural selection.

PLANTS OF THE PAST

Geologists estimate that the earth is about 4½ billion years old, and it seems likely that life has existed on the earth for about 3 billion years. What information we have about the life of past geological ages has been derived from the fossil record, and, as we have already noted, this record is far from complete. Practically no plant or animal fossils older than 600 million years are known, although there is indirect evidence that life already existed about 2½ billion years before that time.

We shall consider briefly the kinds of plants probably present in each of the five eras into which geologists have divided the history of the earth. Geological eras are subdivided into periods, and in most cases it will be necessary to follow the course of plant evolution period by period.

Archeozoic Era. The Archeozoic Era ended about a billion and a half years ago and extended back to the beginning of the earth, about 4½ billion years ago. No fossils are known from the rocks of this era, but sedimentary limestone over one billion years is believed to have been formed by lime-secreting algae. The level of evolution attained by organisms in the next era is another of the bits of evidence

The Evolution of Plants

that has led biologists to conclude that life originated during the Archeozoic. Life was undoubtedly restricted to the oceans and adjacent tidal basins. The simple plants and animals of this era probably lacked structures that would fossilize well; furthermore, the Archeozoic rocks have undergone extensive modification and distortion which would destroy what fossils might have been formed, so the absence of fossils is not surprising.

Proterozoic Era. This era began with the end of the Archeozoic and continued for about a billion years, until about 600 million years ago. Deposits of lime, graphite, and iron indicate the presence of algae and bacteria similar to those that now form such deposits, and fossil bacteria and algae from the Proterozoic have been found recently. Animal fossils include those of simple worms and crustaceans, indicating that the evolution of invertebrate animals had already made considerable progress. Plants were no doubt restricted to algae, bacteria, and perhaps some simple fungi, and life was still restricted to the oceans.

Paleozoic Era. The Paleozoic Era began about 600 million years ago and ended about 230 million years ago, lasting about 370 million years. Although it was shorter than either of the two previous eras evolution proceeded far and for the first time a good fossil record was left. During the first two periods of the Paleozoic (the Cambrian and Ordovician) few plant fossils were left, but there are fossil algae of modern types. Animal fossils include those of all phyla of invertebrates, and even simple vertebrates had evolved. Life was still restricted to the oceans, but plants may have begun to invade the land by the end of the Ordovician. The first direct fossil evidence of land plants comes from the next period (the Silurian), but most of the fossils of this period are those of marine algae. The land fossil resembles the present day club mosses.

The Devonian period, which began about 405 million years ago and lasted until 345 million years ago, was the time when land plants really became well established. The earliest Devonian fossils of land plants are those of psilophytes, primitive vascular plants with creeping stems and erect branches, but no leaves or roots. Mosses and liverworts may have invaded the land earlier, but their soft tissues were not likely to be fossilized. The ancient psilophytes became extinct by the end of the Devonian, but two related genera still survive (Chapter 1). It is likely that a considerable amount of evolution of land

plants occurred during the Silurian without leaving a fossil record, for the land plants of the Devonian were already quite diverse, including primitive club mosses, horsetails, and ferns as well as psilophytes. Some of the species of club mosses were of tree size. Animals were still principally found in oceans, and fish such as sharks and lungfish had evolved, but land animals included scorpions and primitive amphibians.

The Carboniferous period, from about 345 to 280 million years ago, was the time of great forests of club mosses, horsetails, and seed ferns. It was from these treelike plants that most of our coal deposits came, giving rise to the name of the period. Primitive gymnosperms appeared for the first time in the Carboniferous. The land animals included primitive insects and reptiles. The Carboniferous species extended into the next period (the Permian), but during the Permian the climate became drier and perhaps colder and most of the Carboniferous species became extinct, their descendants being in general much smaller plants.

Thus, by the end of the Paleozoic Era all of the major groups of plants except the angiosperms had evolved, even though the gymnosperms had made only a modest start. However, the species present during the Paleozoic were different from our present-day species.

Mesozoic Era. The Mesozoic Era is sometimes called "The Age of Reptiles" since it was during this era that the giant reptiles lived and reptiles were the dominant form of animal life on land. The first birds and mammals evolved during the Mesozoic. The Mesozoic began 230 million years ago and ended 63 million years ago. During the first period of the Mesozoic (the Triassic) the gymnosperms were the dominant land plants, cycads, ginkgo, and conifers becoming very abundant, while the seed ferns became extinct. The gymnosperms retained their dominance during the next period (the Jurassic), but the more primitive species became extinct. Fossil remains indicate a possibility that the first known angiosperms appeared during the Jurassic in the form of primitive woody species. During the last period of the Mesozoic (the Cretaceous) angiosperms gradually became dominant and the gymnosperms began to decline both in abundance and number of species. The Cretaceous angiosperms were woody plants, including familiar genera such as oaks, maples, magnolia, sassafras, willow, and holly.

Cenozoic Era. The Cenozoic Era began 63 million

years ago and is still in progress. During the early Cenozoic many of the modern families of angiosperms became well established, though only wood species apparently existed. Extensive forests covered the earth. Later in the period herbaceous species began appearing and gradually became a more and more important part of the vegetation, while forests dwindled and many tree species became extinct. As the Cenozoic progressed the flora became more and more as it is at present with the evolution of present-day species and the extinction of some of the older ones. The modern birds and mammals evolved during the Cenozoic, man appearing only during the last million years.

Of course, remnants of the once more-abundant gymnosperms, ferns, club mosses, and horsetails survived and gave rise to our present-day species, and mosses and liverworts and the ancient algae and fungi continued their evolution to the present-day forms.

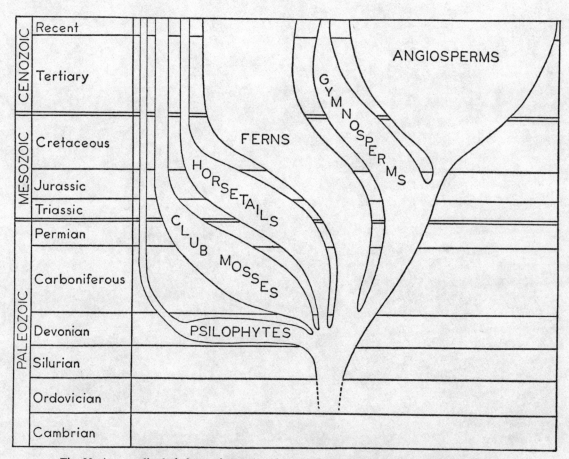

Fig. 99. A generalized phylogenetic tree based on the fossil record and illustrating one possible course of evolution of the vascular plants. The width of the spaces representing each group are more or less proportional to the relative abundance of each group in each of the geological periods, but angiosperms are presently more abundant than indicated by the width of the space.

Both the vascular plants and bryophytes probably evolved independently from the algae. All the major groups of algae and fungi (though perhaps few of the present species, if any) had probably evolved by early Paleozoic times, but the scanty fossil record of non-vascular plants makes it impossible to chart them on a geological time scale as has been done for the vascular plants. The Archeozoic and Proterozoic Eras are not included in this chart. (*Modified from an exhibit at the Chicago Museum of Natural History.*)

The Evolution of Plants

There are too many missing links for the reconstruction of the pathway of evolution from one type of plant to another with any degree of certainty. However, evidence from both the fossil record and the structure of present-day species makes it possible to construct reasonable evolutionary trees, indicating probable relationships and origins. One such scheme is presented in Fig. 99. At any rate, there is little doubt but that our present-day species of plants have evolved through some such pathway, whether or not all the details have been worked out correctly.

Chapter 15

RELATIONS BETWEEN MAN AND PLANTS

Many people are quite unaware of plants and practically never consider their great importance in human life, while others simply take plants for granted as part of the scheme of things. However, mankind as a whole has long been interested in plants from one standpoint or another and many people make their living dealing with plants or plant products.

From time to time in the previous chapters we have referred to various interrelations between plants and man, and Chapter 1 was devoted to a consideration of the plant families most important economically. In this chapter we shall summarize the relations we have already mentioned and consider some additional ones. For convenience we may classify the relations as biological, economic, aesthetic, and scientific, though there is a great amount of overlapping among these categories.

BIOLOGICAL RELATIONS

Here we are dealing with those fundamental and essential biological interrelations that existed before as well as after the dawn of civilization, primarily the dependence of man (as of other animals) on plants as the ultimate source of all food as well as the oxygen of the air. We are dealing, too, with the other interactions between plants and man as members of biological communities, such as human diseases caused by bacteria, fungi, or other plants, plant poisoning, and allergies caused by pollen or mold spores. Forests and other plant communities have provided suitable habitats for man, and man has had an immense influence on natural plant communities as he has cleared land for cultivation or habitations.

ECONOMIC RELATIONS

Plants and plant products constitute one of the major segments of our commerce and industry. We are dependent on plants not only for our biologically essential food but also to a great extent for the other two basic essentials of civilized life: clothing and shelter. Our supply of energy used in heating our homes, operating our automobiles and other means of transportation, operating our various electrical appliances, and running our factories comes primarily from fuels (coal, oil, gas), and these contain energy from the sun trapped by photosynthesis millions of years ago.

While food, clothing, shelter, and fuel constitute our most essential needs, many other important items of commerce are derived from plants. To name but a few, there are antibiotics and other valuable drugs, paper, cork, rubber, rope, wood, alcohol, and tobacco. Aside from the many people who work in farms, gardens, or forests producing these varied items, many others work in factories that process them and still others are concerned with their distribution to consumers. Plants are undoubtedly big business.

AESTHETIC RELATIONS

The beauty of many plants and their flowers has attracted most people, and indeed this is the only standpoint from which many people are really conscious of plants, or are interested in them. Most people with yards around their homes devote considerable time, effort, and money to the cultivation of ornamental plants and the landscaping of their lawns. The extensive raising of houseplants is another aspect of our aesthetic interest in plants, while the provision of flowers for bouquets and arrangements by florists is a big business, as is the provision of ornamental plants by nurseries. The attractiveness of public parks and buildings is greatly increased by proper landscaping, and some of the newer super-highways are very attractive because of good land-scaping and the prohibition of billboards. Botanical gardens probably appeal to more people from the aesthetic standpoint than the scientific one.

Many of our desirable natural landscapes have been ruined by the bulldozer, as they have been invaded by highways, buildings, or cultivation, but some of the more attractive areas have been saved by becoming parts of our systems of state or national parks or forests. The attractiveness of these areas from both a scenic and recreational standpoint generally depends to a great extent on their natural vegetation. Even the clear streams and lakes of an unspoiled area depend on the presence of a vegetational cover that prevents the soil from eroding into the streams.

In addition to our direct aesthetic appreciation of plants, we have used plants extensively as subjects and inspirations for various arts. Flowers, fruits, and other plant parts have been the subject of many still-life paintings and plants play a major role in most landscape paintings. Architectural embellishments, as on the capitals of columns, are frequently plant designs. Many textile and wallpaper designs involve plants or flowers. Plants frequently appear in poems, and have provided the titles if not the inspiration for various musical compositions.

SCIENTIFIC RELATIONS

Man's scientific interest in plants is concerned primarily not with the beauty or economic importance of plants but with the obtaining of a better understanding of the nature of plants—how they are constructed and carry on their life processes, how they grow, develop, reproduce, and transmit their traits to offspring, and how they interact with other living things. Such scientific study of plants is the province of botany, and this book has been devoted to outlining some of the more important scientific facts and principles about plants that have been re-

vealed by botanical research. Botanical research, like research in the other pure sciences, is interested in revealing new facts and principles whether or not they may be of any practical value. However, research in applied botany is directed toward discovering ways of producing more and better plants for man's use.

Among the branches of applied botany are horticulture, agronomy, soils, forestry, plant breeding, and plant pathology. **Horticulture** deals with garden and orchard crops, some of its subdivisions being floriculture (ornamental plants), olericulture (garden vegetable crops), and pomology (orchard fruits). **Agronomy** deals with field crops such as corn, wheat, tobacco, and cotton. **Plant breeders** and **plant pathologists** may specialize in any of these groups of plants or may deal with forest trees, the plant pathologists playing a very important role in controlling and treating the numerous diseases of economic plants. Specialists in all these areas are highly trained, most of them holding a Ph.D. degree.

In general, the aim of applied botany is to provide plants of the best possible heredity by scientific plant breeding, and then to provide the best possible environment for them. This involves making full use of the knowledge accumulated by pure botanical research as well as that obtained by research in applied botany, particularly in areas of plant physiology such as plant growth substances, mineral nutrition, and plant metabolism.

There are also applications of botany outside the general area of agriculture. Breweries, distilleries, and companies that produce yeast all employ plant physiologists and mycologists (botanists who deal with fungi) to work with the basic problems of yeast culture and fermentations. Drug companies that obtain antibiotics and other useful drugs from molds also employ the same types of botanists as well as plant geneticists who develop superior new strains of molds. There is also a constant search for fungi that may produce previously unknown antibiotics or other drugs. Drug companies also use botanists in searching for new drugs from higher plants.

Molds are also sources of various chemicals, such as citric acid, and companies that produce such chemicals employ mycologists and plant physiologists. Some mycologists specialize in the fungi that cause human diseases and play an important role in the medical sciences, as do botanists who help allergists identify pollen and spores that cause allergies. Of course, bacteriologists play important practical roles in medicine and industry, and in a broad sense they are botanists since they deal with plants, but bacteriology has become pretty well established as a science separate from botany.

As just one more example of a quite different type of applied botany we may consider the role of paleobotany (the science of fossil plants) in the discovery of new deposits of petroleum. Particularly important are the palynologists, who deal with fossilized pollen and spores. These have proved to be particularly good indicators of petroleum.

The relationship of these various kinds of applied botany to pure botany is the same as the relation of engineering and industrial chemistry to pure chemistry, or of petroleum or mining geology to pure geology. Specialists in applied botany may be engaged in development or production as well as in applied research, just as engineers may be, but they should be clearly distinguished from operators such as farmers, gardeners, greenhouse workers, practicing foresters, or the men who actually culture yeasts or molds in industry. Such people who deal with plants are generally neither pure nor applied botanists, though they commonly make use of available scientific knowledge in their work with plants. They may be regarded as practicing an applied art rather than as scientists, and of course they play very essential roles in our economy.

Returning to pure botany, we find that it can be subdivided into various specialties on two different bases: from the standpoint of the kinds of plants studied or from the standpoint of the aspect of plant life studied.

On the first basis we have **mycology,** the study of fungi; **bacteriology; algology** (or **phycology**), the study of algae; **bryology,** the study of mosses and liverworts; and **dendrology,** the study of trees. From the standpoint of approach we have plant **anatomy** and **morphology,** the study of structure and form; plant **physiology,** the study of life processes; plant **taxonomy,** the study of classification and relationships; plant **ecology,** the study of plants in relation to their environment; plant **genetics,** the study of heredity; plant **cytology,** the study of cells; plant **phylogeny,** the study of evolution; and **paleobotany,** the study of fossil plants.

The two ways of subdividing botany interact with one another, producing a large number of specialties based on both the kind of plant studied and the way in which it is studied. Thus, a plant morphologist may deal with vascular plants or he may specialize in some group of lower plants and thus may be a bryologist, mycologist, or algologist, for example. Botanists who deal primarily with vascular plants generally are identified just by their approach, while those who deal with non-vascular plants are also identified (or primarily identified) by the kind of plant studied.

Botany is, of course, one of the biological sciences that deals with living organisms, in contrast with the physical sciences such as physics, chemistry, geology, and astronomy. While it is convenient to classify the natural sciences into these various disciplines, it should be made clear that the various sciences are quite closely interrelated with one another. Thus, the biological sciences (botany, zoology, bacteriology, and related fields) are quite dependent on the physical sciences, particularly chemistry and physics, as biological phenomena are basically chemical and physical in nature. In turn, all the natural sciences are quite dependent on mathematics. Biomathematics is now emerging as a new specialty that cuts across the lines dividing the sciences, just as biochemistry and biophysics have done in the past.

Within the biological sciences the dividing lines have also become rather indistinct in many cases, especially in such areas as genetics, ecology, biochemistry, and some aspects of physiology where the same general principles apply to both plants and animals. Many specialists in such areas work with either plants or animals, whichever may be more suitable material for certain experiments, and regard themselves as biologists rather than either botanists or zoologists.

OCCUPATIONS DEALING WITH PLANTS

There are, then, many different occupations in which people deal with plants in various ways and at various levels. First, we have the pure botanists who study plants scientifically and who generally hold Ph.D. degrees in the area of their specialty. Botanists may be members of college or university faculties, devoting varying portions of their time to teaching and research, or they may devote their time to research at U. S. Department of Agriculture or state experimental stations, in industrial laboratories, in research institutes such as the Boyce Thompson Institute for Plant Research at Yonkers, New York, or in the major botanical gardens. Botanists at botanical gardens may also devote considerable time to the botanical education of the public. Then there are the various kinds of applied botanists, who generally also have the Ph.D. degree. They, too, may teach and do research in colleges and universities, but many of them work in federal or state experimental stations or in industrial laboratories. Another highly trained profession dealing with plants is landscape architecture, but this is an art rather than an applied science, the approach being from the aesthetic standpoint.

At a semiprofessional or subprofessional level are the technicians and research assistants who help both the pure and applied botanists in their work. Some of these hold a bachelor's or master's degree in botany or a related science but others have had little or no formal training and have learned their work on an apprentice basis. There is a need for more people who like to engage in this type of work.

The largest occupational group dealing with plants consists of those who cultivate plants—farmers, gardeners, greenhouse operators, and seed company and nursery employees. Modern methods of agriculture have reduced the need for farmers, but there is a great demand for capable people in the other categories. While many people engaged in these occupations have had no special training other than that acquired by experience, there is a need for people who have some technical training and some knowledge of the facts about plants revealed by modern research in pure and applied botany. Of course, there are still other occupations dealing with plants such as florists and tree surgeons. Those who like plants and enjoy working with them can find occupational opportunities at just about any level of training.

BOTANICAL HOBBIES

Many people have derived much pleasure and profit from hobbies dealing with plants. Perhaps the most numerous are the gardeners, who may specialize either in vegetables or flowers, frequently further specializing in some particular kind of plant such as orchids, lilies, roses, iris, or chrysanthemums. There are societies, devoted to the culture and study of each of these and other groups of plants, composed mostly of amateurs who have become quite expert

in the group. Many other people collect and identify different kinds of native plants, particularly wild flowers, trees, grasses, mushrooms, or specific kinds of algae such as diatoms. Sometimes these amateur botanists become quite expert in their chosen group of plants and may be of considerable assistance to professional botanists. Still other people like to experiment with plants by investigating the influence of various plant growth substances, growing plants by hydroponics, or practicing different means of grafting and plant propagation.

Some people have become actively interested in the conservation of our natural resources and find considerable pleasure and satisfaction in working with organizations such as the Nature Conservancy and the Wild Flower Preservation Society.

Those who are interested in some botanical hobby can secure information and assistance from a botanical garden or a botany department of a college or university.

THE FUTURE OF BOTANY

Dr. Robert Oppenheimer, the famous physicist, once made the statement that the past half century has belonged to the physical sciences but that the coming half century will belong to the biological sciences. There is little doubt but that botany and the other biological sciences are developing rapidly and that new discoveries in the coming years will provide us with a much better understanding of plants and animals than we have now, particularly with respect to their heredity, their life processes, and their growth and development. At least some of the discoveries are likely to be spectacular. A young person interested in a career in science may well find the opportunities in the biological sciences greater than in the physical sciences, and will certainly find botany or another biological science an active and interesting field in which to work.

Although botanical research is not directed primarily toward practical goals, it is quite likely that many of the coming discoveries will be found useful in providing us with economic plants of greatly improved heredity and in enabling us to control plant growth and development so as to greatly improve both quantity and quality of plant products. At least two of the major problems facing man are to a large extent biological and their solution is bound to be helped by increased botanical knowledge and application of the knowledge we already have. One is the conservation of our natural resources and the other is the provision of an adequate supply of food for the rapidly increasing human population of the earth. In addition, if space travel develops as it promises, we face another problem that must be solved by the application of botanical knowledge—the provision of both an adequate supply of food and an adequate supply of oxygen for the space travelers. Research on this problem is already under way, and it appears that unicellular algae may provide the answer.

GLOSSARY

Abscission (ab-SIZH-un)—The detachment and falling of leaves or other plant structures.

Absorption—1. The entry of water, mineral salts, and other substances into plants. 2. The trapping of light or other radiant energy by a substance. 3. In general, the taking in or soaking up of one substance by another.

Accumulation—1. The amassing of foods in a plant when the rate of food production (by photosynthesis) exceeds the rate of food use (by respiration and assimilation). 2. The continued entry of ions into cells against a diffusion pressure gradient, resulting in their being more concentrated in the cell than outside. This process occurs at the expense of energy from respiration.

Adenosine-triphosphate (a-DEN-oh-seen tri-FOS-fate)—A complicated organic compound made from an organic base (adenine), ribose sugar, and three phosphate groups. The third phosphate is readily removed, releasing energy usable in cellular processes. Adenosine-triphosphate (ATP) thus is converted to adenosine-diphosphate (ADP). The energy needed to convert ADP and phosphate into ATP may be provided by either respiration or photosynthesis.

Adsorption—The condensation and holding on the surface of solid particles of a thin layer of molecules of a liquid, gas, or solute.

Adventitious (ad-ven-TISH-us)—Plant organs such as buds or roots formed at other than their usual locations.

Aerobic (ay-ur-OH-bick)—Requiring the presence of oxygen gas, as from the air.

Agar (AH-gar)—A substance secured from red algae that forms a thick jelly with water and is used as a solidifying agent for nutrient media for culturing microorganisms and tissues.

Alkaloid—One of many different alkaline, nitrogen-containing organic ring compounds produced by some plants, generally bitter, often poisonous, and having marked physiological effects on animals, *e.g.*, nicotine, morphine, quinine, strychnine.

Amino acid (uh-ME-no)—Organic acids containing nitrogen as $-NH_2$. Proteins are built up from amino acids of about twenty different kinds.

Amylase (AM-i-lace)—An enzyme that hydrolyzes (digests) starch to maltose.

Anaerobic (an-air-OH-bick)—Referring to processes that can occur without gaseous oxygen.

Annual—A plant that completes its life cycle in one growing season.

Anther—The upper part of a stamen that produces pollen.

Antheridium (an-ther-ID-ih-yum)—A male sex organ within which sperms are produced.

Anthocyanin (an-thuh-SIY-a-nin)—One of various water-soluble substances produced by plants and commonly present in the vacuoles of cells, generally having a red or blue color.

Antibiosis—The production by organisms of substances (antibiotics) that inhibit the growth of other organisms.

Apical dominance—The suppression of growth of lateral buds by the terminal bud of a stem.

Archegonium (ar-ki-GO-nee-um)—A female sex organ of a plant containing an egg.

Ascospore (AS-ka-spore)—One of the spores formed within an ascus and resulting, first from sexual fusion, and then, from meiosis.

Ascus (AS-kus)—The saclike sporangium of the ascomycete fungi within which generally four or eight ascospores are produced.

Asexual (ay-SEK-shoo-ul)—Without sex; referring to reproduction not involving the fusion of gametes.

Assimilation—The formation of protoplasm and cell walls from foods.

ATP—See adenosine-triphosphate.

Autotrophic (auto-TROF-ik)—Referring to organisms that manufacture their own food by photosynthesis or chemosynthesis.

Auxin (AUK-sin)—A class of plant growth hormones characterized by their capacity for promoting cell enlargement and causing a negative curvature in the *Avena* coleoptile test.

Axil—The angle between a stem and the upper side of a leaf.

Bark—The tissues outside the vascular cambium of a woody stem or root and consisting of phloem, cork, and cork cambium.

Basidiospore (buh-SID-i-o-spore)—One of the four spores borne externally on a basidium and produced following sexual fusion and then meiosis.

Basidium—The characteristic club-shaped or filamentous sporangium of the basidiomycete fungi.

Bast fiber—Phloem or pericycle fibers, particularly abundant in flax, hemp, and linden and used in making fabrics.

Biennial—A plant that normally lives only two growing seasons and then dies, growing vegetatively the first year and producing flowers only the second year.

Bisexual—1. Applied to an organism that produces both eggs and sperms. 2. Applied to a flower bearing both functional stamens and pistils.

Blade—The broad, flat portion of a leaf.

Bolting—The elongation of internodes of a stem which had previously been in a rosette condition, resulting in a tall stem and generally brought about by low temperature preconditioning, long photoperiods, or a combination of the two.

Bract—A modified leaf, generally smaller than the ordinary foliage leaves and occurring just below a flower or inflorescence.

Bud—1. An embryonic stem tip bearing young leaves (leaf bud), one or more flowers (flower bud), or both leaves and flowers (mixed bud). 2. A smaller cell produced by the division of a parent yeast cell and remaining attached to the parent cell for some time before separation.

Budding—1. The vegetative reproductive process of yeasts, involving the production of buds. 2. A type of grafting in which a bud with a piece of the bark surrounding it is placed under a slit in the bark of the tree or shrub to which it is being grafted.

Bud scale—One of the modified leaves that covers the buds of most woody plants.

Bulb—A short, conical stem bearing several concentric layers of fleshy modified leaves, as in onion or tulip; essentially a large bud containing considerable accumulated food and functioning in vegetative propagation.

Burning—1. Killing of plants by plasmolysis, commonly by use of too much fertilizer. 2. Oxidation of organic compounds with the rapid release of heat and light (flame), in contrast with the controlled, stepwise oxidation of organic compounds in respiration.

Callus—A rather unorganized and undifferentiated parenchyma tissue sometimes formed at or below a wounded plant surface, as a result of plant hormone applications, or in plant tissue cultures.

Calorie—The quantity of heat required to raise the temperature of water one degree centigrade;

also used as a general energy unit. A large or **kilogram calorie** is 1000 of these small or **gram** calories and is the unit used in nutrition and in much biological work.

Calyx (KAY-liks)—The undermost series of flower parts, composed of sepals, which are usually green and leaflike but may be colored.

Cambium (CAM-be-um)—The cylinder of meristematic cells, one cell thick, located between the xylem and phloem of many stems and roots, that gives rise to the secondary xylem and phloem by continuing cell division. The cork cambium (phellogen) gives rise to cork cells.

Capsule—1. A term applied to the sporangium of mosses and liverworts. 2. A dry, dehiscent fruit that has developed from a compound pistil.

Carbohydrate—A large class of organic compounds including starch, cellulose, and sugars, composed of carbon, hydrogen, and oxygen and with the ratio of hydrogen to oxygen generally being 2:1. More technically, carbohydrates are aldehyde or ketone derivatives of polyhydroxy alcohols.

Carnivorous—Feeding upon animals.

Carotenoids (ka-ROT-eh-noids)—A class of yellow, orange, or occasionally red pigments that are fat-soluble and found in the chloroplasts and chromoplasts of plants, and including carotenes and xanthophylls.

Carpel—The megasporophylls of seed plants, each carpel bearing ovules on its margins. A pistil is composed of one or more carpels.

Catalyst (KAT-a-list)—A substance that influences the rate of a chemical reaction (commonly greatly accelerating it) but that is not consumed in the reaction. Enzymes are organic catalysts produced by organisms.

Cell—The basic structural unit of plant and animal bodies consisting of an organized protoplast with a membrane, and, in most plant cells, surrounded by a cell wall.

Cell sap—The water solution of salts, sugars, and other substances occupying the vacuoles of cells.

Cellulose—A fibrous carbohydrate composed of many glucose units linked together and insoluble in water or other common solvents; the principal constituent of most plant cell walls.

Chemosynthesis—The production of foods from carbon dioxide and water by certain species of bacteria by the use of energy obtained from the oxidation of inorganic compounds.

Chlorenchyma (klo-RENG-keye-mah)—1. A parenchyma cell containing chloroplasts. 2. A tissue composed of chlorenchyma cells.

Chlorophyll—One of several chemically related green pigments (chlorophyll a, b, etc.) found in plant cells (generally in plastids) and functioning in absorbing light energy used in photosynthesis.

Chloroplast—A plastid containing chlorophyll. Carotenoids are also present.

Chlorosis (klo-ROW-sis)—A deficiency in chlorophyll, resulting in pale green or yellow areas that are normally green.

Chromatid (KROW-ma-tid)—One of the two threads resulting from the duplication of a chromosome before separation. After separation each chromatid is referred to as a chromosome.

Chromoplast—A plastid containing carotinoids but not chlorophyll.

Chromosome—Small bodies, generally thread-shaped or rod-shaped, present in the nucleus of a cell and containing the genes or hereditary potentialities, which are composed of deoxyribonucleic acids (DNA).

Class—A related group of orders in the scheme of classification.

Climax—The terminal community in a succession. A climax community reproduces itself more or less unchanged until there is a major environmental change.

Clone (klone)—All the offspring of a plant that have been produced by vegetative propagation.

Coenocytic (see-no-SIT-ick)—A cell, filament, or hypha having more than one nucleus not separated by cell walls.

Coenzyme (ko-EN-zime)—A nonprotein organic compound that plays an essential role in various reactions activated by enzymes.

Cohesion—The attractive force that holds together the molecules of a substance.

Coleoptile (koh-le-OPP-till)—A sheathlike portion of a grass cotyledon that covers the young leaves of the seedling.

Collenchyma (kuh-LENG-kuh-muh)—A parenchyma type of cell with extensive secondary thickening of the cell walls at the corners only.

Colony—1. A cluster of cells, attached to each other but independent of one another and commonly all alike, as in some algae. 2. A mass of cells derived generally from a single cell ancestor, as in bacteria.

Companion cell—A small parenchyma-type cell always found attached to the side of a sieve tube element in the phloem of angiosperms.

Compound leaf—A leaf divided into two or more leaflets.

Conidium (koe-NID-ih-um)—An asexual fungus spore produced singly or in chains at the end of a hypha rather than in a sporangium.

Cork—A secondary tissue commonly found in the bark of woody plants and composed of closely arranged cells that at maturity are dead and have walls waterproofed with waxy and fatty materials (suberin).

Corm—A short, broad, fleshy underground stem with a vertical axis, as in the gladiolus.

Corolla (kuh-RAHL-uh)—The second lowest series of flower parts, composed of petals.

Cortex—The primary tissues (generally parenchyma, chlorenchyma, or collenchyma) of a stem or root, located between the epidermis and the phloem or endodermis.

Cotyledon (kotl-EE-d'n)—A seed leaf of an embryo plant.

Crossing over—The change in linkage of the genes of a linkage group, brought about by the interchange of parts between the chromosomes of a homologous pair.

Crown—The head of foliage of a tree or shrub.

Cuticle—The layer of waxy and fatty substances covering the epidermal cells of stems and leaves.

Cutin (CUE-tin)—The substance of which a cuticle is composed.

Cytoplasm (SIY-tuh-plasm)—The parts of a protoplast except for the nucleus.

Deciduous—Dropping of all leaves of a tree or shrub at the end of a growing season.

Deoxyribonucleic acid—The nucleic acid of chromosomes that constitutes the genes; characterized by its content of deoxyribose sugar and four kinds of organic bases—ademine, cytosine, guanine, and thymine—as well as phosphate units; commonly called DNA.

Diffusion—The movement of the molecules or other particles of a substance under their own kinetic energy from a region of higher diffusion pressure to a region of lower diffusion pressure of that particular substance.

Diffusion pressure—The total molecular activity of a substance as determined by its concentration, velocity of the molecules (influenced by temperature), and imposed pressure.

Digestion—The enzymatic hydrolysis of complex and usually insoluble foods into simple and soluble foods.

Dioecious (die-EE-shus)—A plant bearing both staminate and pistillate flowers.

Diphosphopyridine nucleotide—A complex organic compound that functions as a coenzyme of dehydrogenase enzymes; referred to as DPN.

Diploid—Having two sets (the $2n$ number) of chromosomes per cell.

Disaccharide (die-SACK-uh-ride)—A sugar made from two molecules of simple sugars (or monosaccharides).

DNA—Deoxyribonucleic acid.

Dominant—1. A gene that expresses itself even though another different gene of the same set is present. 2. The most conspicuous or characteristic species of an ecological community.

Dormancy—A lack of growth of buds or seeds resulting from some internal inhibiting factor rather than from an unfavorable environment.

DPN—Diphosphopyridine nucleotide.

Dry Weight—The weight of a plant tissue or soil after all its water content has been evaporated by heating.

Egg—A female gamete.

Embryo sac—The female gametophyte of angiosperms.

Endodermis—A one-cell-thick tissue of roots (and rarely stems) between the cortex and pericycle.

Endoplasmic reticulum—Branched tubular or plate-like membranes extending throughout the cytoplasm of a cell and containing the ribosomes.

Endosperm—A tissue surrounding an embryo in many kinds of seeds and containing much accumulated food. The endosperm of angiosperms arises from the fusion of the two polar nuclei of the embryo sac and a sperm and is triploid, while the endosperm of gymnosperm seeds arises from the tissue of the female gametophyte and is haploid.

Energy—The capacity for doing work. Energy may be kinetic, electrical, radiant (light, etc.), heat, chemical, etc.

Enzyme—A protein catalyst produced by an organism that increases the rate of a specific biochemical reaction.

Epidermis—The outermost layer of cells of leaves, young stems, and young roots.

Epiphyte (EP-uh-fite)—A plant that grows attached to another plant but that does not secure food from it, *i.e.*, is not parasitic on it.

Etiolation (EE-tih-oh-LAY-shun)—The combination or characteristics of a plant that has been growing in the dark or in dim light, *e.g.*, long, slender stems, small leaves, and deficiency of chlorophyll.

Family—The unit of classification larger than a genus but smaller than an order; often used, but incorrectly, for the different species of a genus or even the varieties of a species.

Fat—An organic compound made from glycerol and fatty acids; one of the principal kinds of food.

Fatty acid—An organic acid generally having a long carbon chain and only one carboxyl (—COOH) or acid group at the end.

Fermentation—1. Anaerobic respiration. Oxygen is not used and pyruvic acid produced by glycolysis is converted to substances such as alcohol or lactic acid rather than proceeding through the Krebs Cycle and the terminal oxidations to water and carbon dioxide; particularly characteristic of yeasts and bacteria of various species. 2. In a broader sense, any process carried on by microorganisms that results in the production of quantities of alcohols, acids, or other organic compounds, whether anaerobic or not, as in acetic acid fermentation.

Fertilization—1. The union of two gametes with the formation of a zygote. 2. The process of applying fertilizer to plants.

Fertilizer—A substance containing one or more essential mineral elements and supplied to plants to avoid or correct a deficiency of these elements.

Fiber—Greatly elongated, thick-walled supporting cells pointed at both ends; present in pericycle, phloem, and the xylem or wood of angiosperms.

Field capacity—The percentage water content of a soil when it holds all the capillary water it can, but no gravitational water.

Filament—1. A threadlike chain of cells, as in the bodies of many species of algae. 2. The stalk of a stamen.

Fission—The reproduction of a unicellular organism by division into two cells of equal size.

Flagellum—A threadlike protuberance of a cell that moves in a whiplike fashion and causes movement of the cell.

Flora—The native plants of a particular region at present or in some past time.

Florigen—A hypothetical plant hormone produced by leaves and causing the formation of flower buds, formed in many species only during photoperiods of a certain range of durations.

Flower—A determinate, sporophyll-bearing stem axis.

Food—Any organic substance that can be used by an organism as a source of energy (in respiration) or as material for building its cells (assimilation); principally carbohydrates, fats, and proteins.

Free space—The parts of a cell through which substances may readily pass by diffusion or flow. It includes most parts of the cell except for the vacuoles, plastids, mitochondria, nuclei, and perhaps the ribosomes.

Frond—A leaf of a fern.

Fruit—The ripened ovulary of angiosperms within which seeds are borne, together with adjacent structures that may be fused with the ripened ovulary.

Gall—Abnormal outgrowths from plants caused by the presence in the tissues of the larvae of gall wasps, other insects, fungi, or bacteria, having characteristic and often elaborate structure.

Gamete—A cell capable of fusing with another cell in the process of sexual reproduction, producing a zygote that can grow into a new individual; gametes may be eggs, sperms, or isogametes.

Gene—The unit of inheritance, located in the chromosomes of the nucleus; genes are now believed to be composed of deoxyribonucleic acid (DNA).

Genera—Plural of genus.

Genotype—The sum total of the genes of an organism, or of such of its genes as are being considered in a particular genetic investigation.

Genus (JEE-nus)—In classification, a group of closely related species, *e.g.*, the various species of oaks.

Geotropism (jee-OTT-ruh-pizm)—Growth movements in response to gravity.

Germination—Resumption of growth of a seed or spore.

Gibberellin (jibb-er-ELL-in)—A plant growth substance that causes marked elongation of internodes and in certain species also promotes blooming and other development.

Girdling—Removal of a complete ring of bark from the stem of a tree or shrub.

Glycolysis (glye-KOLL-uh-sis)—The initial series of reactions in both aerobic and anaerobic respiration that result in the conversion of sugar to pyruvic acid with the release of a small amount of energy.

Grain—The fruit of members of the grass family, consisting of a single seed tightly surrounded by the fruit, as in wheat and corn; incorrectly referred to as a seed.

Granum—A stack of thin discs, present in considerable numbers in chloroplasts, containing the chlorophylls and carotenoids and serving as the site of the light reactions of photosynthesis in higher plants.

Gravitational water—Water that drains through a soil by the force of gravity.

Growth substance—A plant growth hormone or synthetic substance that influences plant growth and development markedly at low concentrations.

Guard cell—A specialized and usually bean-shaped epidermal cell that is a component of a stomate and that with its mate determines the size of the stomatal opening between them by changes in shape.

Guttation (guh-TAY-shun)—The loss of liquid water from the ends of leaf veins, the drops being forced out under pressure.

Gymnosperm (JIM-no-sperm)—The group of seed plants in which the seeds are not enclosed within an ovulary and thus have no fruits.

Habitat—The natural environment in which a plant usually grows.

Haploid—The presence of only one set of chromosomes in the cells of an organism.

Hardwood—The wood of angiosperm trees, as contrasted with the softwood of gymnosperm trees; also applied to any tree having hardwood.

Haustorium (haws-TOR-ih-yum)—A specialized outgrowth of a parasitic fungus or vascular plant (*e.g.*, dodder) that penetrates the tissue of the host and absorbs food, water, and other substances from it.

Heartwood—The older and more central rings of wood that contain no living cells and are generally darkened as compared with the younger and outer sapwood.

Herb—A non-woody plant.

Heterosporous (het-er-OS-por-us)—Having spores of two different kinds, *i.e.*, microspores that develop into male gametophytes and megaspores that develop into female gametophytes.

Heterotrophic (het-er-o-TROFF-ik)—Organisms that obtain their food from outside sources, *i.e.*, those that cannot make their own food by photosynthesis or chemosynthesis.

Heterozygous (het-er-o-ZYE-gus)—The situation wherein the two genes of a pair in a diploid organism are unlike.

Hexose—A simple sugar (monosaccharide) with six carbon atoms.

Homosporous (hoe-MOS-per-us)—Having but one kind of meiospore.

Homozygous (hoe-moe-ZYE-gus)—The situation wherein the two genes of a pair in a diploid organism are alike.

Hormone—An organic substance produced in one part of an organism and transported to other parts where it exerts marked influences on growth or other processes in low concentrations.

Hummock—A small hill or rounded mound.

Humus—The portion of a soil composed of a mixture of partially decomposed organic matter.

Hybrid—1. Offspring of a cross between two species. 2. Offspring of a cross between two members of a species differing in one or more of the pairs of genes under consideration.

Hydrolysis (hiy-DROL-ih-sis)—A chemical reaction in which water reacts with another substance, causing the latter to break down into still other substances, *e.g.*, digestion of foods.

Hydrophyte (HIY-druh-fite)—A plant that grows partially or completely submerged in water.

Hydroponics (hiy-druh-PON-iks)—The culturing of

plants in solutions of essential mineral elements, rather than in soil.

Hypha (HIY-fuh; pl. hyphae (HIY-fee)—The thread-like filaments that make up the body of a fungus.

Hypocotyl (hiy-puh-KOT-l)—The part of a seedling or embryo plant between the cotyledons and the radicle.

Imbibition (im-bih-BISH-un)—Diffusion of water into a colloidal solid (*e.g.*, wood) and its absorption on internal surfaces, causing the substance to swell.

Inflorescence (in-flaw-RESS-ens)—A flower cluster; the characteristic arrangement of flowers on the stem.

Integument (in-TEG-you-ment)—The one or two outer layers of an ovule that develop into the seed coats.

Interfascicular (in-ter-fuh-SICK-you-ler)—Between vascular bundles.

Internode—The region of a stem between two successive nodes.

Ion—A portion of a molecule bearing an electrical charge.

Isogamy (eye-SOG-uh-me)—A type of sexual reproduction in which all the gametes (isogametes) are of the same size and shape, rather than differentiated as eggs and sperms.

Isotope—Any of the several forms of a chemical element having the same chemical properties and atomic number but different atomic weights. The different isotopes of an element all have the same number of protons and electrons but different numbers of neutrons.

Kinetic energy—Energy of movement.

Kinetin (KEYE-nuh-tin)—A plant growth substance essential for cell division and having certain other influences on plants.

Lamella—A thin plate, sheet, or layer.

Latex—A milky fluid found in certain species of plants.

Leaf gap—An interruption of the vascular tissue of a stem above the point at which a bundle of vascular tissue (the leaf trace) entered a leaf from the stem.

Leaflet—One of the several individual blades of a compound leaf.

Legume—A member of the pea or bean family (*Leguminosae*).

Lenticel—Small areas of loosely arranged cork cells on the surfaces of roots and stems through which gases can diffuse.

Leucoplast (LOO-kuh-plast)—A colorless plastid.

Lichen (LIE-ken)—A composite organism consist-ing of unicellular algae characteristically located among the hyphae of a fungus.

Lignin—A complex organic compound that is insoluble and quite unreactive chemically; the principal constituent, except for cellulose, of the secondary cell walls of fibers, vessel elements, tracheids.

Linkage—The failure of two genes to assort independently, resulting from the fact that they are located on the same chromosome.

Lipid (LIH-pid)—Any of a class of chemical compounds formed from fatty acids and alcohols that are insoluble in water but soluble in fat solvents, *e.g.*, fats, oils, waxes.

Liter (LEE-ter)—A metric unit of volume equal to 1.0567 quart; 1000 cubic centimeters (milliliters).

Lumen (LOO-men)—The region inside the cell wall of a dead cell, such as a fiber, occupied by water or air and formerly occupied by the protoplast.

Marine—Pertaining to the sea or ocean.

Megaspore—In heterosporous species, the spore that develops into the female gametophyte.

Meiosis (mye-OWE-sis)—The two successive cell divisions (of the spore mother cell in plants) resulting in reduction of the chromosome number from the diploid to the haploid number.

Meiospore (MYE-owe-spore)—One of the four spores resulting from each meiotic division.

Meristem—A tissue composed of actively dividing young cells that gives rise to the various tissues of a plant by the enlargement and differentiation of some of the older meristematic cells.

Mesophyll (MES-uh-fil)—The chlorenchyma tissues between the upper and lower epidermis of a leaf, including both palisade and spongy layers.

Mesophyte—A plant that grows under conditions of medium moisture, *e.g.*, most forest plants.

Metabolism—The sum total of the biochemical and biophysical processes of an organism.

Micron—A metric measure of length, equal to 1/1000 millimeter.

Micropyle—The small porelike opening through the integuments of an ovule.

Microsome—A very small spherical body found in numbers on the endoplasmic reticulum of a cell and composed of ribonucleic acid (RNA); presumably the site of protein synthesis; synonym for ribosome.

Microspore—In heterosporous plants, a spore that develops into a male gametophyte.

Microtome—An instrument for use in cutting very thin sections of tissue for microscopic examination.

Middle lamella—A thin layer of pectic substances common to two adjacent cells and serving as an adhesive that holds the cells together; considered as a part of the cell walls and located between the adjacent cellulose walls.

Mitochondrion (mye-toe-CON-drih-yon, pl. mitochondria)—A minute cytoplasmic body, commonly oval or rod-shaped, presumably the site of the Krebs Cycle and terminal oxidation parts of respiration.

Mitosis (my-TOE-sis)—Division of a nucleus by a sequence of processes that results in each of the two nuclei formed having the same number and kinds of chromosomes as the original nucleus; mitosis is usually followed by cytokinesis, *i.e.*, the division of the cytoplasm of cell division.

Monocotyledon—An angiosperm plant that has only one cotyledon, *e.g.*, grasses, lilies; also, monocot.

Monoecious (muh-NEE-shus)—A situation in which one plant bears both staminate and pistillate flowers; not applicable to plants having perfect flowers bearing both pistils and stamens in each flower.

Monosaccharide—A simple sugar, *i.e.*, one that cannot be digested into smaller and simpler sugar molecules.

Mutation—A change in a gene or chromosome resulting in the appearance of a new and different hereditary potentiality.

Mutualism—The living together of two organisms of different species with each organism benefiting from the interrelationship.

Mycelium (mye-SEE-li-um)—A mass of fungus hyphae, particularly when compact as in mushroom fruiting bodies.

Mycorhiza (my-co-RYE-zuh, pl. mycorhyzae)—A mutualistic structure composed of roots covered by or invaded by the hyphae of certain species of fungi and apparently functioning in absorption.

Nitrification—The conversion of ammonia nitrogen to nitrites and then nitrates by two kinds of nitrifying bacteria, which use the energy from this oxidation in making foods by chemosynthesis.

Nitrogen fixation—The conversion of the gaseous nitrogen of the atmosphere into nitrogen compounds.

Node—A point on a stem where a leaf and its axillary bud are borne.

Nucleic acid—Complex organic compound with large molecules built up from four different organic bases, ribose sugar, and phosphate; see deoxyribonucleic acid and ribonucleic acid.

Nucleolus (new-klee-OWE-lus)—The spherical bodies found within a nucleus.

Nucleoprotein—A protein linked with a nucleic acid.

Nucleus—The relatively large cell structure, commonly spherical or ovoid, containing the chromosomes.

Ontogeny (on-TODJ-eh-nee)—The developmental life history of an organism, from zygote to maturity.

Order—A category of classification into which a class is divided and in turn composed of a group of related families.

Organ—A structure composed of a group of organized tissues and constituting one of the principal structural components of an organism; in vascular plants the organs are basically roots, stems, and leaves.

Organelle—A specialized cell structure.

Organic compound—A chemical compound containing carbon and hydrogen. Oxygen and other elements are found in some organic compounds.

Osmosis—The diffusion of water through a differentially permeable membrane.

Ovary—A female sex organ that produces one or more eggs; commonly but improperly applied to the ovulary of a pistil; archegonia are really ovaries, but are not referred to as such.

Ovulary—The expanded and hollow basal portion of a pistil containing ovules.

Ovule—The structures within the ovularies of pistils that develop into seeds.

Oxidation—An energy-releasing chemical reaction, generally involving the addition of oxygen or removal of hydrogen from a compound.

Parasite—An organism that lives in or on another organism (the host) and secures its food from the host.

Parenchyma (puh-RENG-ki-muh)—A type of cell with thin walls and large central vacuole that is living when mature; a tissue composed of such cells.

Parthenocarpy—Formation of fruits without pollination and fertilization.

Pathogen—A parasite that causes a disease.

Pectin—A complex organic compound present in the middle lamellae and primary walls of cells and contributing the jelling power to fruit juices.

Pedical—The stalk of an individual flower in an inflorescence.

Pentose—A simple sugar containing five carbon atoms per molecule.

Perennial—A plant normally living more than one or two years and not dying after one blooming period.

Perianth—The sepals and petals of a flower, generally used when both are the same colors.

Pericycle (PER-ih-siy-kul)—The parenchyma tissue

between the endodermis and the phloem of a root.

Petiole—A leaf stalk.

Phenotype (FEE-no-type)—The visible characters of an organism in contrast with the hereditary potentialities of its genotype.

Phloem (FLOW-em)—The vascular tissue containing sieve cells or sieve tubes and functioning in the translocation of foods and other solutes.

Photoperiodism—The response of plants to the relative lengths of day and night.

Photosynthesis—The production of sugar (or perhaps other foods) and oxygen from carbon dioxide and water by the use of light energy absorbed by chlorophyll.

Phototropism (foe-TOT-ro-pizm)—Bending growth resulting from unequal illumination.

Phylogeny (fye-LODJ-ih-nee)—The evolutionary history of a species or larger classification group.

Phytochrome—A plant pigment existing in two interconvertible forms, one absorbing primarily red light and the other primarily longer wave red (far red) light; present in low concentration but important in photoperiodism and several other growth and development processes.

Pistil—The flower organ composed of one or more carpels and containing ovules.

Pit—A hole or recess in the secondary walls of a plant cell.

Pith—The tissue occupying the center of a stem and usually made up of parenchyma cells.

Plankton—The aggregation of small plants and animals floating at or just below the surface of a body of water.

Plasma membrane—The cytoplasmic membrane adjacent to the wall of a cell.

Plasmodesm—A minute thread of cytoplasm connecting the protoplasts of adjacent cells through a small opening in the walls.

Plasmodium (plaz-MOE-di-um)—The vegetative stage in the life cycle of a slime mold, consisting of a multinucleate mass of protoplasm without any cell walls.

Plasmolysis (plaz-MOLL-uh-sis)—Shrinking of the protoplasm of a cell away from the cell wall as a result of the diffusion of water out of the vacuole.

Plastid—Any of several kinds of specialized cytoplasmic bodies; *e.g.*, chloroplasts, chromoplasts, leucoplasts.

Plumule (PLOO-mule)—The terminal bud of an embryo or young seedling.

Polyploid—An organism whose cells contain more than two sets of chromosomes.

Polysaccharide—A carbohydrate whose molecules are built up from many molecules of simple sugars and is commonly neither sweet nor soluble, *e.g.*, starch, cellulose.

Protein—A class of organic compounds with large molecules built up from many molecules of amino acids; one of the three principal classes of foods and the principal organic constituent of protoplasm.

Protonema (pro-toe-NEE-ma)—The young gametophyte plant of a moss, commonly consisting of a green filament resembling an algal filament.

Protoplasm—Living matter, composed principally of water, proteins, carbohydrates, lipids, and nucleic acids, and highly organized, generally into nucleus, cytoplasm, and their various component structures.

Protoplast—The protoplasm of a cell.

Rachis (RAY-kis)—The main stalk of a compound leaf to which the leaflets are attached.

Radicle—The lower end of an embryo that gives rise to the primary root.

Ray—Vertical ribbon of parenchyma extending radially through the vascular tissue. Pith rays originate and extend from the pith outward, while wood rays originate in any of the annual rings of the xylem.

Receptacle—The part of a modified stem (peduncle) that bears the flower parts.

Reproduction—The detachment from an organism of cells or groups of cells that can grow into new individuals, either by themselves or after fusing with other cells.

Respiration—The oxidation processes within a cell whereby the chemical energy of foods is released in a form usable in doing work in the cell.

Rhizoid (RYE-zoid)—A filamentous and sometimes branched extension from the body of a nonvascular plant or the gametophyte of a vascular plant that functions in absorption.

Rhizome—An underground stem that is usually horizontal in position and frequently woody or fleshy.

Ribonucleic acid—A type of nucleic acid found in the ribosomes as well as the nucleus of a cell, differing in composition from deoxyribonucleic acid in that it is made from ribose sugar rather than deoxyribose and contains uracil rather than thymine; called RNA.

Ribosome—See Microsome.

RNA—Ribonucleic acid.

Root cap—The thimblelike parenchyma covering the root apex.

Root hair—Long tubular outgrowths from root epidermal cells found in a zone above the region of cell elongation.

Root pressure—The pressure developed in roots that may cause water to flow upward through the xylem and that is responsible for guttation and exudation.

Rosette—A dense cluster of leaves growing close to the ground, resulting from the fact that the internodes of the stem bearing them have not elongated.

Runner—A stolon.

Saprophyte (SAP-row-fite)—A plant that uses non-living organic matter as food.

Scion (SIY-un)—The stem or bud grafted onto the roots and lower stem (the stock) of another plant.

Sclerenchyma (skli-RENG-kih-muh)—Thick-walled, supporting cells that are not living at maturity, either long and pointed at both ends (fibers) or about the same diameter in each direction (sclerids or stone cells); or a tissue composed of such cells.

Seed—An embryo plant enclosed in seed coats and, in many species, an endosperm located between the embryo and seed coats; a matured ovule.

Seed coat—The outer tissue of a seed, derived from the integument of the ovule.

Sepal (SEE-pal)—One of the lower cycle of flower parts (the calyx), often green and leaflike but colored like the petals in some species.

Sessile—A leaf, flower, or fruit without a stalk.

Shoot—1. The upper part of a plant consisting of the stem and the leaves, flowers, and fruits it bears. 2. A young branch that is actively growing.

Shoot tension—The tension or negative pressure developed in the cells of leaves or other parts of a shoot that causes water to flow upward through the xylem and serves as the principal means of water translocation in vascular plants; syn. cohesion mechanism, transpiration stream.

Shrub—A relatively low woody perennial with several main stems or branches arising from or near the ground, in contrast with the single main stem or trunk of a tree.

Sieve cell—A type of living cell, found in the phloem of gymnosperms, having perforated walls and serving in the translocation of foods and other solutes.

Sieve tube—A vertical series of conducting cells (sieve tube elements) found in the phloem of angiosperms.

Sieve tube element—One of the cells constituting a sieve tube and characterized by lack of a nucleus when mature and the presence of sievelike perforations in the walls, particularly at the ends of the cells.

Sol—A colloid in a liquid condition, as contrasted with a colloidal gel.

Solute—A dissolved substance; the discontinuous phase of a solution.

Solvent—A substance (generally liquid) in which other substances are dissolved; the continuous phase of a solution.

Species—The basic unit of classification, consisting of a group of organisms having many characteristics in common and generally capable of interbreeding freely; species may, however, be subdivided into subspecies, varieties, races, or strains.

Sperm—A male gamete.

Sporangium—A structure within which spores are produced.

Spore—Any of several kinds of minute and usually unicellular reproductive bodies produced by plants and that develop into new plants without fusing with another cell. True **asexual spores** are produced only by algae and fungi. **Meiospores** are a stage in the sexual life cycle of plants, are produced from the spore mother cells of sporophytes by meiosis, and develop into gametophyte plants. **Resting spores** have resistant walls and are generally a means of survival through unfavorable periods rather than means of reproduction. **Zygospores** are zygotes formed from isogametes and functioning as resting spores. **Zoospores** are flagellated spores.

Spore mother cell—A diploid mother cell that produces haploid meiospores (usually four) by meiosis.

Sporophyll (SPOR-uh-fil)—A modified leaf that has sporangia within which meiospores are produced.

Sporophyte (SPOR-uh-fite)—The diploid generation of a plant life cycle that reproduces by means of meiospores.

Stamen (STAY-men)—The flower part that produces pollen; the microsporophyll of angiosperms.

Starch—A polysaccharide carbohydrate synthesized from glucose by plants and that is neither sweet nor soluble but forms grains.

Stele—The cylinder or core of tissues of a root (or less commonly a stem) surrounded by the endodermis and including the pericycle, phloem, xylem, and pith.

Stigma—The tip of a pistil, generally expanded and sticky, that receives pollen.

Stipe—A stemlike structure that is not a true stem, *e.g.*, the stalk of a mushroom or kelp.

Stipule—A small bladelike leaf part found at the base of the petiole of leaves of some species.

Stock—The root and lower stem of a plant to which another plant is grafted.

Stolon—1. A horizontal and usually aboveground branch stem that roots at its nodes; *i.e.*, a runner.

2. A horizontal aerial hypha of a fungus, *e.g.*, bread mold.

Stomate (STOW-mate)—A pair of guard cells and the closeable pore between them, found in the epidermis of vascular plants. Also stoma (pl. stomata).

Style—The part of the pistil between the stigma and ovulary.

Suberin (SUE-ber-in)—The waxy substances present in the walls of cork cells.

Substrate—A substance acted upon by an enzyme in a biochemical reaction.

Succession—The orderly sequence of plant communities in an area, terminating in a climax community that then occupies the area until there is a major climatic or geological change or disturbance by man.

Symbiosis (sim-bye-OH-sis)—1. The close ecological association between organisms of two or more species. The association may be harmful to one of the organisms (*e.g.*, parasitism) or may be beneficial to both of them (mutualism). 2. In a narrower sense, mutualism, as specified under Definition 1.

Synapsis—The pairing of homologous chromosomes during meiosis.

Synthesis—A chemical reaction in which a substance is built up from one or more simpler substances.

Teleology (tee-le-OL-o-ji)—The explanation of natural phenomena on the basis of purposeful behavior.

Tetrad—A group of four meiospores produced from a single spore mother cell by meiosis.

Tetraploid—Having the 4*n* number of chromosomes, or four sets, in a cell, in contrast with the more usual diploid or 2*n* number.

Thallus—A plant body lacking roots, stems, and leaves, as in algae, fungi, and bryophytes, as well as in the gametophytes of vascular plants.

Thermoperiodism—The influence of the daily or seasonal fluctuations in temperature on plant growth, development and behavior.

Tissue—An organized group of cells that are generally adjacent to one another and are identical or similar in origin, structure, or function.

TPN—Triphosphopyridine nucleotide.

Tracheid (TRAY-ki-id)—An elongated cell tapering at both ends with thick pitted walls and a lumen; found in the xylem, particularly of gymnosperms, and serving in support and water transport.

Translocation—The flow of water and of solutions of mineral salts, foods, hormones, and other substances through the vascular tissue of plants; sometimes referred to as conduction.

Transpiration—The loss of water vapor from plants.

Triose—A sugar containing three carbon atoms per molecule.

Triphosphopyridine nucleotide—A complex organic compound that serves as a hydrogen acceptor in photosynthesis and to some extent in respiration.

Tropism (TROWE-pizm)—A bending growth movement in response to an environmental factor (*e.g.*, light, gravity) that is of unequal magnitude on two sides of the bending organ.

Tuber—A much enlarged and fleshy portion of an underground stem.

Turgor—The swollen or inflated condition of a cell resulting from the diffusion of water into the cell.

Turgor pressure—The pressure created by the diffusion of water into a cell and exerted against the cell wall.

Unicellular—Composed of or having only one cell.

Vacuole (VACK-you-ole)—A droplet of water within the cytoplasm of a cell and containing sugars, salts, pigments, and other substances dissolved in water. This solution occupying the vacuole is called the cell sap.

Vascular tissue—The xylem and phloem, *i.e.*, the conducting or translocation tissues.

Vegetable—1. Any plant. 2. Any part of a plant (root, stem, leaf, flower, fruit, or seed) commonly served as part of the main or salad course of a meal.

Vegetation—The plant communities of the earth or a particular region.

Vegetative—Referring to a process or structure related to maintenance of the individual, in contrast with reproductive processes or structures related to the maintenance of the species.

Vein—A vascular bundle of a leaf blade.

Vernalization—The influence of low temperatures on the development and blooming of a plant at a later date; low temperature preconditioning.

Vessel—A long (one inch to several feet) tube in the xylem or wood of angiosperms composed of a longitudinal series of dead cells with their end walls partially or entirely disintegrated, through which water flows.

Vessel element—One of the cells from which a vessel is formed.

Viable—Living; commonly applied to seeds.

Virus—One of various minute, submicroscopic bodies composed of nucleoproteins and sometimes other substances that invade living organisms and are duplicated by them, commonly causing various plant or animal diseases.

Vitamin—Any of various complex organic compounds (not chemically related to one another) synthesized by plants and essential in minute quantities for the normal functioning and growth of animals and (in most cases, at least) of plants and microorganisms. Vitamins function catalytically, generally as coenzymes or components of coenzymes, and in the plants that produce them they generally have all of the characteristics of plant hormones.

Whorl—1. Three or more leaves or branches at a node. 2. A circle of flower parts, *e.g.*, petals or stamens.

Wilting—The loss of turgidity by the leaves on stems and their conesquent dropping, as a result of a deficiency of water. **Temporary wilting** results from a rate of transpiration higher than the rate of water absorption, even though there is adequate soil water. **Permanent wilting** occurs because of the depletion of all available soil water.

Wilting percentage—The percentage water content

of a soil at the time permanent wilting occurs.

Wood—The xylem; the part of the vascular tissue that contains vessels or tracheids and transports water in vascular plants; applied in particular to the xylem of trees and shrubs.

Xanthophyll (ZAN-tho-fill)—One of various yellow, orange, or red carotenoid pigments found with carotenes in plastids and differing from carotenes in oxygen content.

Xerophyte (ZEE-row-fite)—A plant that grows in deserts or other environments having a limited water supply.

Xylem (ZYE-lem)—The vascular tissue that contains tracheids or vessels and conducts water; wood.

Zoospore (ZOE-oe-spore)—A motile spore that swims by means of whiplike flagella.

Zygote (ZEYE-goat)—The cell resulting from the fusion of two gametes.

FOR FURTHER READING

The following selected list of books will no doubt include a number that you will find interesting and profitable once you have acquired a general knowledge of botany by reading *Botany Made Simple*. Most of the books in the list are non-technical and have been written with the general reader rather than the specialist in mind.

The list does not include two general classes of books about plants in which you may be interested: manuals for the identification of various groups of plants such as trees, wild flowers, ferns, mushrooms, and algae, and books on various aspects of gardening. These were omitted because the best selections vary from one region of the country to another. Recommendations as to the best books for your region can be obtained from botanists at nearby colleges and universities, or in the case of garden books from your state agricultural experiment station. The experiment station can also provide you with a list of its free bulletins on gardening, while a list of U. S. Department of Agriculture bulletins dealing with gardening and other aspects of plant life can be obtained from the U. S. Government Printing Office, Washington, D.C. 20401. Botanical gardens and museums are also good sources of inexpensive bulletins dealing with plants.

You may also wish to keep up-to-date on various developments in botany through magazine articles. The two magazines best for this are the *Scientific American* and *Natural History,* each of them publishing many botanical articles. Reprints of articles in which you are particularly interested can generally be obtained without cost by writing to the author. The W. H. Freeman Publishing Company, San Francisco, California, has made special reprints of many *Scientific American* articles in the various fields of science and sells them for twenty cents each. A free catalog of the reprints is available.

Two of the paperback books of articles reprinted from the *Scientific American* and published by Simon and Schuster include articles on plants: *Plant Life* and *The Physics and Chemistry of Life.* These articles go into some detail on many topics that could be treated only briefly in *Botany Made Simple,* and you may wish to add them to your library.

Other magazines that publish articles on botanical subjects from time to time include the *Saturday Evening Post, Harper's,* the *Reader's Digest,* and *Science Digest. Science News Letter* is a weekly publication containing news items and short articles on recent advances in the various fields of science. Of course, there are the various gardening magazines and these sometimes include articles on botanical developments as well as practical gardening articles. The English quarterly *Endeavour* publishes excellent articles on developments in the various sciences including botany, but they are generally a little more technical than those in the *Scientific American* and *Natural History.* Most libraries have all the magazines mentioned. You might find it interesting to go through the back volumes of the *Scientific American* and *Natural History* looking for articles on subjects you would like to know more about.

BAKER, H. G. *Plants and Civilization.* Belmont, Calif.: Wadsworth Publishing Co., 1965. A paperback on economic botany.

BAKER, J. J. W. AND ALLEN, G. E. *Matter, Energy, and Life.* Palo Alto, Calif.: Addison-Wesley, 1965. A paperback dealing with the chemical background of biology.

BELL, C. R. *Plant Variation and Classification.* Belmont, Calif.: Wadsworth Publishing Co., 1967. A paperback.

BILLINGS, W. D. *Plants and the Ecosystem.* Belmont, Calif.: Wadsworth Publishing Co., 1965. A paperback on plant ecology.

BOLD, HAROLD C. *The Plant Kingdom.* Englewood Cliffs, N.J.: Prentice-Hall, 1964. A paperback that surveys the various groups of plants.

BONNER, DAVID AND MILLS, S. E. *Heredity.* Englewood Cliffs, N.J.: Prentice-Hall, 1964. An up-to-date paperback introduction to genetics.

CHRISTENSEN, C. M. *The Molds and Man.* Minneapolis: University of Minnesota Press, 1951.

CRONQUIST, ARTHUR. *Introductory Botany.* New York: Harper & Row, 1961. A college botany textbook that is particularly good on the various groups of plants, plant structure, and plant evolution.

DARRAH, W. C. *Principles of Paleobotany*. New York: The Ronald Press Co., 1960. A textbook on fossil plants that is not too technical and is easy to read.

FAIRCHILD, DAVID. *The World Was My Garden*. New York: Charles Scribner's Sons, 1954. An interesting account of the author's explorations throughout the world for new plants of economic value.

GABRIEL, M. L. AND FOGEL, S. *Great Experiments in Biology*. Englewood Cliffs, N.J.: Prentice-Hall, 1955. Reprints of important biological research papers going back to the Middle Ages.

GALSTON, A. E. *The Life of the Green Plant*. Englewood Cliffs, N.J.: Prentice-Hall, 1964. A paperback introduction to plant physiology, particularly good on plant growth.

GREULACH, VICTOR A. AND ADAMS, J. EDISON. *Plants: An Introduction to Modern Botany*. New York: John Wiley & Sons, 2nd ed., 1967. A college textbook particularly good on plant physiology, ecology, and genetics.

HANAUER, ETHEL R. *Biology Made Simple*. Garden City, N.Y.: Doubleday & Co., 1956.

HARTMAN, H. T. AND KESTER, D. E. *Plant Propagation*. Englewood Cliffs, N.J.: Prentice-Hall, 1959.

HESS, FRED C. *Chemistry Made Simple*. Garden City, N.Y.: Doubleday & Co., 1955. Recommended as background reading for Botany Made Simple for those who have never taken a chemistry course.

JENSEN, W. A. *The Plant Cell*. Belmont, Calif.: Wadsworth Publishing Co., 1965. A paperback.

KEOSIAN, J. *The Origin of Life*. New York: Reinhold Publishing Co., 1964. A paperback.

KNOBLOCH, I. W. *Selected Botanical Papers*. Englewood Cliffs, N.J.: Prentice-Hall, 1963. A collection of classical and current research papers and articles on botany.

MC ELROY, W. D. *Cellular Physiology and Biochemistry*. Englewood Cliffs, N.J.: Prentice-Hall, 1964. A paperback.

MITCHELL, J. W. AND MARTH, P. *Growth Regulators in Field, Garden and Orchard*. Chicago: University of Chicago Press, 1947.

SCHERY, R. W. *Plants for Man*. Englewood Cliffs, N.J.: Prentice-Hall, 1960. An economic botany text.

SEARS, PAUL B. *Deserts on the March*. Norman: University of Oklahoma Press, 1935. A readable book on conservation from the viewpoint of a plant ecologist, with special reference to dust storms.

———. *Where There is Life*. New York: Dell Publishing Co., 1962. A paperback on ecology and conservation.

STORER, J. H. *The Web of Life: A First Book on Ecology*. New York: The Macmillan Co., 1959.

TORREY, J. G. *Development in Flowering Plants*. New York: The Macmillan Co., 1967. A paperback.

VEEN, R. VAN DER AND MEIJER, G. *Light and Plant Growth*. New York: The Macmillan Co., 1959.

INDEX